AF729871

Intelligent Systems, Control and Automation: Science and Engineering

Volume 102

Series Editor

Kimon P. Valavanis, Department of Electrical and Computer Engineering, University of Denver, Denver, CO, USA

Advisory Editors

P. Antsaklis, University of Notre Dame, Notre Dame, IN, USA

P. Borne, Ecole Centrale de Lille, France

R. Carelli, Universidad Nacional de San Juan, Argentina

T. Fukuda, Nagoya University, Japan

N. R. Gans, The University of Texas at Dallas, Richardson, TX, USA

F. Harashima, University of Tokyo, Japan

P. Martinet, Ecole Centrale de Nantes, France

S. Monaco, University La Sapienza, Rome, Italy

R. R. Negenborn, Delft University of Technology, The Netherlands

António Pascoal, Institute for Systems and Robotics, Lisbon, Portugal

G. Schmidt, Technical University of Munich, Germany

T. M. Sobh, University of Bridgeport, CT, USA

C. Tzafestas, National Technical University of Athens, Greece

Intelligent Systems, Control and Automation: Science and Engineering book series publishes books on scientific, engineering, and technological developments in this interesting field that borders on so many disciplines and has so many practical applications: human-like biomechanics, industrial robotics, mobile robotics, service and social robotics, humanoid robotics, mechatronics, intelligent control, industrial process control, power systems control, industrial and office automation, unmanned aviation systems, teleoperation systems, energy systems, transportation systems, driverless cars, human-robot interaction, computer and control engineering, but also computational intelligence, neural networks, fuzzy systems, genetic algorithms, neurofuzzy systems and control, nonlinear dynamics and control, and of course adaptive, complex and self-organizing systems. This wide range of topics, approaches, perspectives and applications is reflected in a large readership of researchers and practitioners in various fields, as well as graduate students who want to learn more on a given subject.

The series has received an enthusiastic acceptance by the scientific and engineering community, and is continuously receiving an increasing number of high-quality proposals from both academia and industry. The current Series Editor is Kimon Valavanis, University of Denver, Colorado, USA. He is assisted by an Editorial Advisory Board who help to select the most interesting and cutting edge manuscripts for the series:

Panos Antsaklis, University of Notre Dame, USA
Stjepan Bogdan, University of Zagreb, Croatia
Alexandre Brandao, UFV, Brazil
Giorgio Guglieri, Politecnico di Torino, Italy
Kostas Kyriakopoulos, National Technical University of Athens, Greece
Rogelio Lozano, University of Technology of Compiegne, France
Anibal Ollero, University of Seville, Spain
Hai-Long Pei, South China University of Technology, China
Tarek Sobh, University of Bridgeport, USA

Springer and Professor Valavanis welcome book ideas from authors. Potential authors who wish to submit a book proposal should contact Thomas Ditzinger (thomas.ditzinger@springer.com)

Indexed by SCOPUS, zbMATH, SCImago.

Maria Isabel Aldinhas Ferreira ·
Mohammad Osman Tokhi
Editors

Towards Trustworthy Artificial Intelligent Systems

Springer

Editors
Maria Isabel Aldinhas Ferreira
Centro de Filosofia
Universidade de Lisboa
Lisbon, Portugal

Mohammad Osman Tokhi
School of Engineering
London South Bank University
London, UK

ISSN 2213-8986 ISSN 2213-8994 (electronic)
Intelligent Systems, Control and Automation: Science and Engineering
ISBN 978-3-031-09822-2 ISBN 978-3-031-09823-9 (eBook)
https://doi.org/10.1007/978-3-031-09823-9

© The Editor(s) (if applicable) and The Author(s), under exclusive license to Springer Nature Switzerland AG 2022
This work is subject to copyright. All rights are solely and exclusively licensed by the Publisher, whether the whole or part of the material is concerned, specifically the rights of translation, reprinting, reuse of illustrations, recitation, broadcasting, reproduction on microfilms or in any other physical way, and transmission or information storage and retrieval, electronic adaptation, computer software, or by similar or dissimilar methodology now known or hereafter developed.
The use of general descriptive names, registered names, trademarks, service marks, etc. in this publication does not imply, even in the absence of a specific statement, that such names are exempt from the relevant protective laws and regulations and therefore free for general use.
The publisher, the authors, and the editors are safe to assume that the advice and information in this book are believed to be true and accurate at the date of publication. Neither the publisher nor the authors or the editors give a warranty, expressed or implied, with respect to the material contained herein or for any errors or omissions that may have been made. The publisher remains neutral with regard to jurisdictional claims in published maps and institutional affiliations.

This Springer imprint is published by the registered company Springer Nature Switzerland AG
The registered company address is: Gewerbestrasse 11, 6330 Cham, Switzerland

Preface

> *"People have little reason to turn away from machines; which are nothing other than hyperdeveloped and hyperconcentrated forms of certain aspects of human subjectivity [...] it will be possible to build a two-way bridge between human beings and machines and, once we have established that, to herald new and confident alliances between them".*
>
> *Felix Guatari*

Artificial intelligent systems, in either embodied or non-embodied forms, have been increasingly permeating multiple domains in the economic, social and cultural environments that characterize the present lived reality. And though this reality may be distinct depending on the regional context, the human intertwining with technology is a horizontal phenomenon common to all.

Innovation is driven by the very evolution of society and by its goal to address the needs of its members fostering wealth and individual and collective physical and psychological well-being. This goal is actually present in the 3 Ds motto that for decades has been driving robotic research and development: to free human beings from dangerous, dirty and dull tasks.

Millions of end-users, of consumers have been responding positively to the technological development first made possible by electricity and mass production, in the early twentieth century, happily adopting the lifestyle changers that became the average consumer dreams: a new TV set, a huge refrigerator, a dishwashing machine, the latest car version, new services... and envisioning the robotic butler that would perform all the dull tasks in the domestic universe.

On its side, industry and business fed and driven by the demands of a very fast-growing market have tried to define the different potential end-users' profiles not only to better respond to their interests and needs but also to foster new interests and new needs, even when these were/are totally superfluous. Consumers have been embracing innovation, incorporating it in their daily lives, considering the objects, the services at their disposal as trustworthy, because they assume them as risk free.

Human-centricity, sustainability and resilience are said to be hallmark features of the present industrial stage[1]. Some of the European Union key policies already reflect this human-centric stance by placing human dignity, human rights, well-being as drivers of technological development and innovation: the General Data Protection Regulation (GDPR)[2] protecting the rights of individuals to their personal data protection; the Machine Directive;[3] and the White Paper on Artificial Intelligence, setting out principles for eventual AI regulation and providing safeguards for the users of certain categories of AI technologies, try to define a legislative framework capable of enforcing this view in the actual practice.

We believe that to be trustworthy technology has to be driven by human needs and shaped according to fundamental human values, creating an inclusive framework where no one is left behind or is segregated. This view has motivated and substantiated the present book that aggregating in a cluster-like way multidisciplinary contributions highlights in a kaleidoscopic perspective how AIS are presently evolving.

In Artificial Intelligence: A Concept Under-Construction, A Reality Under -Development, Ferreira addresses the apparently imprecise semantic substance of the concept [Artificial Intelligence] highlighting the fundamental role played by language in the definition of scientific and technological frameworks. Referring that as technology develops also the concepts it stands on evolve, Ferreira claims that the concept of [Artificial Intelligence] is still a concept under construction. The author tries to identify its boundaries and core semantic substance by contrasting it with the concept it maps on, that of [Natural Intelligence].

What are we referring to when designating a system as trustworthy? This is the theme of the chapter In Machines We Trust?. Ferreira posits that human trust has deep biological roots that resulted (i) from the need to collaborate dictated by the nature of the social species and (ii) the need to be risk aware in order to protect oneself. Stating that in the human sphere trust has both interactive and intersubjective dimensions, Ferreira compares the semantic features that define human/human trust and what it means to trust a product. According to the author, by establishing this boundary, it becomes clear what are the features that define trustworthiness in products, namely, in artificial intelligent systems. Finally, Ferreira addresses the roles of Benchmarking, Standardization and Certification as building blocks of this type of trustworthiness.

Roeland de Bruin, in Informational Privacy and Trust in Autonomous Intelligent Systems, points out that for a successful societal deployment of Autonomous Intelligent Systems (AIS), citizen's trust is crucial. The central theme of this chapter is formed by the relationship between informational privacy protection, trust and acceptance of autonomous intelligent technology. This contribution assumes a legal perspective where certain rules of the EU General Data Protection Regulation (GDPR) are analysed that may apply to AIS-solutions. It is further investigated to

[1] cf, European Commission, TOWARDS SUSTAINABLE EUROPEAN INDUSTRY (2021).

[2] https://gdpr-info.eu/.

[3] The Machinery Directive is the core European legislation regulating products of the mechanical engineering industries.

what extent the application of these rules to AIS-solutions allows citizens to believe that their informational privacy is well protected.[4]

In Ethical Risk Assessment For Social Robots: Case Studies in Smart Robot Toys, Alan Winfield, Anouk van Maris, Katie Winkle, Marina Jirotka, Pericle Salvini, Helena Webb, Arianna Schuler Scott, Jaimie Lee Freeman, Lars Kunze, Petr Slovak and Nikki Theofanopoulou point out the importance of running risk assessment tools in order to produce safe and reliable robots. Emphasizing that risk assessment is a well-known and powerful method for discovering and mitigating risks, and hence improving safety, the authors present an Ethical Risk Assessment methodology that uses the typical ERA approach, but extends the scope of risk to cover ethical risks in addition to safety risks. This chapter outlines the Ethical Risk Assessment framework (ERA) and sets ERA within the broader framework of Responsible Robotics. The use of ERA is illustrated first with a hypothetical smart robot teddy bear (RoboTed), and later with an actual smart robot toy (Purrble). Through these two case studies this chapter demonstrates the value of ERA and how consideration of ethical risks can prompt design changes, resulting in more ethical and sustainable robots.

In Practical and Open Source Best Practices for Ethical Machine Learning, Jeroen Franse, Violeta Misheva and Daniel S. Vale describe the goals of the Foundation for Best Practices in Machine Learning (a non-profit foundation) that seeks to promote responsible ML through creating an open-sourced, freely accessible repository of best practices and associated guides. Its model and organizational guides look at both the technical and institutional requirements needed to promote responsible ML. Blueprints touch on subjects such as "Fairness and Non-Discrimination", "Representativeness and Specification", "Product Traceability", "Explainability" among other topics. Where the organizational guide relates to organization-wide process and responsibilities (i.e. the necessity of setting proper product definitions and risk portfolios); the model guide details issues ranging from cost function specification and optimization to selection function characterization, from disparate impact metrics to local explanations and counterfactuals. It also addresses issues that emerge in the product management.

The Foundation's philosophy is that (a) context is key, (b) responsible ML starts with prudent MLOps and product management and (c) responsible ML needs to be supported by all aspects of an organization's structure.

Selin E. Nugent and Susan Scott-Parker in Recruitment AI has a Disability Problem: Anticipating and Mitigating Unfair Automated Hiring Decisions highlight that Artificial Intelligence (AI) technologies have the potential to dramatically impact the lives and life chances of people with disabilities seeking employment and throughout their career progression. While these systems are marketed as highly capable and objective tools for decision-making, a growing body of research demonstrates a record of inaccurate results as well as inherent disadvantages for historically marginalized groups. Assessments of fairness in Recruitment AI for people with disabilities have thus far received little attention or have been overlooked. In

[4] Assistant Professor AI/law, Centre for Access to and Acceptance of Autonomous Intelligence, Utrecht University; Attorney-at-law (law and technology) KienhuisHoving N.V.

this chapter, the authors examine the impacts to and concerns of disabled employment seekers using AI systems for recruitment and discuss recommendations for the steps employers can take to ensure innovation in recruitment is also fair to all users. Making systems fairer for disabled employment seekers ensures systems are fairer for all.

In Developing and Evaluating Complex Interventions: The Case of Robotic Systems in Cognitive Rehabilitation Therapy, Isabel Ferreira points out that the use of robots in cognitive rehabilitation has been going on, at an experimental level, for decades, comprehending a broad scope of applications ranging from those aiming at the stimulation of patients with dementia to others addressing children in the autism spectrum. Though the results provided by the multiple projects developed in this domain apparently point to the eventual beneficial character of this intervention, its validation and formal recognition in clinical terms still lacks, keeping these experiments punctual and inconsequent from a medical perspective, being the impact of its developments limited to the robotics R&D.

The chapter addresses this issue claiming that the formal clinical recognition of the possible beneficial role of artificial intelligent systems in cognitive rehabilitation therapy is only possible if/when the robotic projects dedicated to this theme are designed and developed as Complex Interventions, i.e. interventions that stand on a lattice of interdependencies and co-effects. According to the Medical Research Council, Complex Interventions are widely used in the health service, in public health practice and in areas of social policy such as education, transport and housing that have important health consequences. The present paper reviews the guidelines and recommendations defined both by the "Framework for Development and Evaluation of Randomized Control Trials (RCTs) for Complex Interventions to Improve Health", April 2000, and in its updated 2019 version "Developing and evaluating complex interventions". Both are intended to help researchers with the design, choosing the appropriate methods, identifying the constraints on evaluation, presenting the evidence in the light of methodological and practical constraints.

The present paper adopts an end-user-centred stance in which the physical and psychological well-being of the patients that are targeted by these interventions, namely, those that volunteer to participate in the trials, as well as that of their families, is viewed as ethically prior. According to this assumption, a fundamental role is given to the monitoring of this well-being and its assessment throughout the intervention process and the monitoring and assessment of the evolution of their health condition before, during the trials and after the experiment, in a long-run mode.

In COVID-19 contact tracing applications in Portugal: effectiveness and privacy issues, Arlindo Oliveira reflects on the Portuguese experience of running contact tracing applications during the COVID-19 pandemics. Referring that Portugal developed and adopted a contact tracing smartphone application, in order to help in the containment of the spreading of the COVID-19 pandemic, the author points out that though the particular technology used guaranteed privacy and security, it did not prove particularly effective, since it was downloaded by only a fraction of the population and very lightly used to report cases of contacts between infected and non-infected persons. The complex issues determining its social acceptance are addressed and discussed in the present text.

Yeh-Liang Hsu in Robots That Look After Grandma? A Gerontechnology Point of View discusses the challenges of using robots for the care of older adults in their homes. Assuming a gerontechnology point of view, the chapter addresses how technology can be accepted as a natural part of older adults' everyday lives at homes.

Developing their reasoning framework based on a case study that takes the famous, heroic ditching of US Airways Flight 1549, an event colloquially known as the "Miracle on the Hudson", Michael Giancola, Selmer Bringsjord, Naveen Sundar Govindarajulu and Carlos Varela, in Making Maximally Ethical Decisions via Cognitive Likelihood and Formal Planning, attempt to give an answer to the following question: Given an obligation and a set of potentially inconsistent, ethically charged beliefs, how can an artificially intelligent agent ensure that its actions maximize the likelihood that the obligation is satisfied? The present approach to answering this question lies in the intersection of several areas of research, including automated planning, reasoning with uncertainty and argumentation.

In The (Uncomputable!) Meaning of Ethically Charged Natural Language, for Robots, and Us, from Hypergraphical Inferential Semantics, Selmer Bringsjord, James Hendler, Naveen Sundar Govindarajulu, Rikhiya Ghosh and Michael Giancola take us to a two-young-child, two-parent household, the Ruben-steins, and their household robot, Rodney, in year 2030. With the parents out, the children ask Rodney to perform some action α that violates a Rubensteinian ethical principle P_R. Rodney replies: (s_1) "Doing that would be (morally) wrong, kids". The argument the children give Rodney in protest is that another household, the Müllers, also has a robot, Ralph; and the kids argue that *he* routinely performs α. As a matter of fact, Ralph's doing α violates no Müllerian ethical principle P_M. Ralph's response to the very same request from the children he tends is: (s_2) "Okay, doing that is (morally) fine, kids". What is the *meaning* of the utterances made by Rodney and Ralph? The authors try to address this question by presenting and employing a novel, formal, inferential theory of meaning in natural language: *hypergraphical inferential semantics* (HIS), which is in the general spirit of **proof-theoretic semantics**, which is in turn antithetical to Montagovian **model-theoretic semantics**, HIS.

In Reinventing Kantian Autonomy for Artificial Agents: Implications for Self–Driving Cars, Endre Kadar and Zsolt Palatinus refer that one of the central problems in Robot Ethics is the possibility of regarding artificial intelligent agents as moral agents. This chapter addresses this problem by proposing to reinvent autonomy for artificial moral agency based on Kantian ethics. In addition to ethics also the Kantian concepts of rationality and laws of freedom are critically discussed. The feasibility of proposal includes examples from existing practices such as human use of animals as autonomous agents and reinterpretation of moral behaviour (rationality, autonomy, freedom, etc.) in autonomous vehicles.

In Realizing the Potential of AI in Africa: It All Turns on Trust, Charity Delmus Alupo, Daniel Omeiza and David Vernon analyse the impact of artificial intelligent systems in developing countries, namely, in Africa. Acknowledging the global disruptive influence of Artificial Intelligence (AI) on all aspects of economy, from manufacturing, to services, to governance but also the potential benefits that embracing AI technologies can bring, the authors point out that the transition from recognition of potential to realization of benefits is not a straightforward matter. In this chapter, they argue that this transition depends on turning technological invention into innovation, that technological innovation cannot happen without adoption and that adoption depends on socio-cultural factors, in general, and on trust, in particular. They conclude by drawing out the implications for AI in developing countries in Africa, arguing that, for Africa to realize the potential of AI in solving economic and social problems, the advancement and deployment of AI must be driven and executed by the peoples of Africa: if it is not, there will be little trust, less adoption and minimal benefits.

Lisbon, Portugal — Maria Isabel Aldinhas Ferreira
London, UK — Mohammad Osman Tokhi

Acknowledgments

The editors would like to thank all the authors who have kindly accepted reflecting on the societal and ethical challenges posed by the deployment of Artificial Intelligent Systems and on the role of shared best practices building trustworthy AIS.

Thank you to Thomas Ditzinger for supporting the initial idea and to all Springer's team for their kindness and professionalism.

Contents

Artificial Intelligence: A Concept Under-Construction, A Reality Under-Development

Maria Isabel Aldinhas Ferreira

Abstract At a time, the term [Artificial Intelligence] increasingly populates public speech blurring the boundaries of its initial scope of reference and running the risk of becoming a general category—an "umbrella term"—both for sociotechnical systems as well as for computational techniques, the present paper aims to grasp its core semantic identity by mapping the concept of [Artificial Intelligence] onto that of its logic correlate [Natural Intelligence]. But the mere definition of [intelligence] is complex as it involves much more than just a "semantic exercise" of deconstruction attempting lexicographic clarification. It demands the definition of an epistemological framework capable of originating and justifying the ontological presence of the term. The present chapter assumes cognition as an embodied, embedded, and always situated process whose dynamo is *semiosis*, i.e., the essential "interpretative" process that takes place within the dialectic relationship binding an organism to its environment. In this epistemological framework, intelligence is viewed not as a human-specific endowment but as an existential attribute inherent of all life forms. An attribute that is responsible for their adequate response to environmental prompts, for their capacity to adapt, their capacity to strive, persist, live and replicate. This framework, that is fundamental to analyse and understand the universal phenomenon of natural cognition and the essence of organic intelligence in its multiple diversity, also provides the necessary grounding to clarify what is at stage when we refer to artificial intelligence, simultaneously allowing to identify the fundamental changes that are being introduced in the typical forms of cognition by the on-going massive incorporation of artificial intelligent systems (AIS) and the digitization of daily reality. We believe this theoretical grounding will contribute (i) to clarify the semantic substance associated to the concept of [Artificial Intelligence] as well as its scope of reference (ii) to identify the new forms of Agent/Environment relationship brought about by the incorporation of sociotechnical systems in different domains of life (iii) to develop (AIS) that are context sensitive, i.e., that are capable of identifying and respecting

M. I. A. Ferreira (✉)
Centre of Philosophy of the University of Lisbon, Lisbon, Portugal
e-mail: isabelferreira@letras.ulisboa.pt

Institute of Systems and Robotics/Instituto Superior Técnico, University of Lisbon, Lisbon, Portugal

© The Author(s), under exclusive license to Springer Nature Switzerland AG 2022
M. I. A. Ferreira and M. O. Tokhi (eds.), *Towards Trustworthy Artificial Intelligent Systems*, Intelligent Systems, Control and Automation: Science and Engineering 102,
https://doi.org/10.1007/978-3-031-09823-9_1

the values substantiated in the social/cultural matrix present in the context they are embedded in, while achieving their expected goals efficiently.

> The beginning of wisdom is the definition of terms
>
> Socrates.

1 A Concept Under-Construction

In the last decades, technological innovation has been progressing at light speed impacting either developed or developing countries.[1] This impactful technological transformation, that is changing the tools, the modes of production and the lifestyles of millions all over the globe is the peak of a complex process of industrialisation that started long ago with the mechanisation of production, made possible by steam powered machinery, and that went on with the massification of this production thanks to electricity, followed by the globalised digital transformation and automation of processes introduced by electronics and information technology, whose exponential computational growth paved the way to the emergence of Artificial Intelligence in its multiple forms.

Probably due to the varied applications in which it can operate plus the distinct domains and contexts in which it can perform, the concept of Artificial Intelligence is presently a broad category whose meaning and ontological nature has become diffuse. Due to the multiple formulations and interpretations proposed for the term,[2] the European Commission attempted a definition to guarantee the needed coherence of the concept. According to this proposal, the term [Artificial Intelligence] designates:

> Any software that is developed with one or more of the techniques and approaches [...] and can, for a given set of human-defined objectives, generate outputs, such as content, predictions, recommendations, or decisions influencing the environments they interact with.

Assigning the correct meaning to a verbal expression is not a mere terminological/linguistic requirement. The meaning assigned to a lexical form, i.e., its semantic substance and corresponding mental representation is a social construct, a value shared by a community of individuals that converges on the identity and interpretation assigned to a given entity.

Ferreira [11, 13] points out that if meanings were just purely individualistic constructs resulting from an act of free-will by the sign-user, language would not fulfil

[1] Following the neo-Kantian approach defined by Ferreira [11, 13].

[2] Cf. Li and Du [32: 1], Haenlein and Kaplan [28], OECD [38: 7], among many others.

its primarily dual function, (i) that of enabling the definition of a personal and collective worldview and (ii) the communication and sharing of experience/knowledge between human beings.

Reflecting on the social character of language, Barthes [3] writes that it is because language is a system of contractual values that it resists the modification coming from a single individual.

The imperative for a community's convergence on the value (meaning) assigned to a term becomes even more urgent in scientific endeavour where language/terminology is the tool that sediments experience allowing for the on-going sharing and construction of knowledge. The temporary conceptual instability that sometimes occurs derives exactly from the "under-construction mode" that typifies research and development processes.

Stuart Glenn refers to this fact in the following terms:

> Concepts, especially as they are employed in the sciences, have a way of moving around. Take any interesting concept found within the sciences – e.g., electricity, force, compound, gene, species, sensation, memory – and you will find that these concepts have histories, histories that are intertwined with empirical research in the fields. [26: 13]

In fact, when looking diachronically at the substance of a concept, we realise that its eventual evolution always reflects how the findings that were withdrawn from empirical observation and/or experiment and that caused theoretical (re-) formulations led to a new conceptual embodiment. This is, in our opinion, the case of the semantic variability registered presently in the use of the term [Artificial Intelligence].

Notwithstanding the instability that is inherent to the under-construction mode of the concept, we believe that a good understanding of its semantic substance and consequently of its scope of reference could be achieved by contrasting [Artificial Intelligence] with the logic correlate he maps into—[Natural Intelligence].

Plus, we believe that by reflecting on the general characteristics of natural cognition and on the role played by natural intelligence in this process, we will be able to get a clearer view of the universal variants involved in each act of cognition and of the fundamental role played by context. This will contribute to the identification of the distinct types of cognition presently observable in human physical and digital environments and the types/forms of agent/environment relationship that embody them. In the case of exclusive artificial cognition or in the case of hybrid modes, this approach will facilitate the development of Algorithmic Sensitiveness to Contextual Variance which is a priority in order to produce artificial intelligent systems capable of identifying the distinct systems of values—socio/cultural matrices—that characterise their different operating frameworks. This awareness of the nature of context will allow for the production of ethically aligned applications that respect the social, cultural and ethical specificities of the framework they are embedded into.

> Nearly always, the organism has been explained on the basis of a preconceived idea of the structure and functioning of the machine; but only rarely have the structure and function of the organism been used to make the construction of the machine more understandable
>
> Georges Canguilhem.

2 Natural Cognition: What Does [Intelligence] Stand for?

2.1 Basic Assumptions

Acknowledging the contributions of a long philosophical inquiry regarding the distinction between living beings and collections of inanimate matter or even man-made mechanisms, Keller [23] points out that a living system is an organism, i.e., "a body that, by virtue of its peculiar and particular organisation, that is not hermetically sealed [...] achieves autonomy and the capacity for self-generation" (ibidem: 2). She goes further stating that whereas a tool requires a tool user, an organism behaves as if it has a mind of its own, that is, it governs itself.

This reflection on the distinctive character of animate entities relatively to non-animate, that some may even find trivial, emerges as prior at a time the production and massive deployment, in all domains of life, of artificial systems exhibiting a certain degree of autonomy is already a fact.

The present chapter, stemming from the epistemological framework defined in Ferreira [11, 13], assumes a neo-Kantian perspective on the nature of cognition, integrating some of the conceptual tools developed by Varela [51], namely the concepts of embodiment, embeddedness and situatedness, and assuming *semiosis* as an essential existential phenomenon, in line with the works of von Uexküll [53, 54], Sebeok [43, 44], Dewey [10], Barbieri [1, 2] and Hoffmeyer [29], among many others.

Taking *semiosis* as the dynamo of cognition,[3] intelligence is here viewed not as a human specific endowment, but as an attribute of all living entities, responsible for their adequate response to environmental prompts, for their capacity to adapt to environmental challenges, their capacity to strive, persist and evolve.

2.2 The Three Common Key Existential Factors

Ferreira [17] posits that human beings, the ultimate social animals, share with all the other life forms three key existential circumstances—embodiment, embeddedness and situatedness Embodiment, i.e. the particular physical architecture different life forms are endowed with, is the joint product of genes and the environmental challenges the system had to face during its evolutionary and developmental history.

[3] In the sense that life itself and the existential dynamics depend on the correctness of this essential "interpretative" phenomenon.

The typology of this physical architecture assumes multiple forms corresponding to different categories and levels of complexity ranging from the reduced single or dual code script of a virus or the “simplicity” of a basic cell to the highly structured network of complex systems that substantiate mammals and in particular human beings.

The essential existential bond that links all these different life forms to their respective environments shows how the coupling of both entities defines a microcosm—a closed purposive organisation.[4] The dynamo of this microcosm is a semiosic process—a process of individuation, identification and meaning assignment—that assumes the form of a dialectic relationship, in which the embedded agent—an entity endowed with a particular physical architecture—and its environment, coupled, interact and evolve by mutually influencing each other.

Cognitive agents embedded in dynamic environments are continually updating their responses to it and this dialectic relationship is responsible for their mutual development and evolution.

In fact, over time, the plastic structures of both the living system and its typical environment change as a result of mutual perturbations. Dewey [10] refers to this dialectic relationship, that emerges from the interaction of organism/environment, as a transaction. A transaction that in Dewey’s terms takes place between two mutually specifying and co-determining systems.

Addressing this co-determination, Thompson et al. [50] give as an example how the trichromatic features that make the bees visual system especially sensitive to ultraviolet seem to have co-evolved with the flowers, present in their environments, and which often display contrasting patterns in the ultraviolet spectrum. This mutually advantageous context seems to be responsible for the co-evolution of the plant’s features and of the sensorial-neural capacities of bees.

We can also observe this mutual co-determination in the mutations viruses undergo in order to optimise their replication potential. In fact viruses evolve far more rapidly than cellular organisms as they are made up of strings of DNA or RNA encapsulated in a protein shell that can only survive and replicate embedded in an adequate environment—the essential metabolic machinery of a living host, which they lack. By consecutively interacting with the hosts’ immune systems and in order to optimise their replication potential viruses undergo changes in their genome. As Fleischemann [22] points out viruses are continuously changing as a result of genetic selection. They undergo subtle genetic changes through mutation and major genetic changes through recombination. Mutation occurs when an error is incorporated in the viral genome. Recombination occurs when co infecting viruses exchange genetic information, creating a novel virus.

von Uexküll [53] coined a fundamental concept—Umwelt. According to this author (ibidem), the term refers to the particular world that each life form defines in its interaction with the environment it is embedded in. Uexkull [54] tries to illustrate the concept by asking the reader to imagine each organism as evolving within an

[4] Cf. on this purpose Gibson [25] concepts of affordances and invariants.

individual soap bubble, pointing out that this imaginary bubble is in fact its existential realm, the only reality of a species, its specific meaningful world. Uexkuhl's soap bubble metaphor is fundamental to (i) fully understand how different physical architectures determine particular worlds and are, on the other hand, also affected by that world and also (ii) to understand the role of a "body-determined intelligence".

This virtual sphere, the figurative perimeter, traced according to the type of interactions allowed by the physical architecture of the organism visualises the embedded character of the agent and the scope of its interactions, by shaping in the general environment the organism's Umwelt, its meaningful existential world.

von Uexküll [53] believes that access to the different worlds of individual life forms can only be reached through the study of their specific organisation. This access demands, according to him, that we submerge ourselves in the anatomical structure of the organism, responsible for defining the way it interacts with the external world, and at the same time that we make as clear as possible the extent of the achievements that this structure brings about, in order to define the field of its existence and its activities.

In "What is it like to be a bat?" [37], Thomas Nagel contemplated his own inability to imagine the world of a life form that navigates the sky by sonar, sending out shrieks at frequencies that the human ear can't even perceive, and locating the presence of trees or insects by means of echoes. But as Ferreira [11] points out it is in this bat-specific Umwelt that particular patterns are perceived as salient, are individuated and are assigned a value—meaning—by the bat, triggering, this way, the agent's adequate behavioural response.

Cassirer [7] points out that this "sensitivity" to particular environmental patterns is neither a predicate attributed to things as such, as absolute things, nor does it consist of simple passive possession of certain sense-data by the organism. The capacity to individuate and identify the environmental patterns the system depends on is determined by its evolved physical architecture and is triggered when the mature system meets the adequate existential context, initiating a life-long "learning process" that rests on its own present and also on its previous lived-experience. Dictated by the requirements imposed by the system's internal state(s), this capacity to individuate and assign meaning to typical environmental features springs out naturally once the adequate coupling conditions are met, c.f. the works of Gottlieb [27], Smotherman and Robinson [47, 48], Lecanuet et al. [31], Lickliter [33], Schaal et al. [40] to mention just a few.

That disposition rests upon a significant degree of innate "knowledge" a "know-how", which all living systems possess and is the result of the experience of biological predecessors and a consequence of their intelligent adaptive efforts to adequately respond to environmental conditions and changes, Ferreira [11].

According to Moreno (1992) ontogenetic adaptation corresponds to the most elemental version of perception as it is achieved through the selective activation of the pertinent genes given the detection of certain environmental conditions and consequently enhancing specific patterns of behaviour.

Addressing the natural evolution of this innateness, Damásio [9] refers that in the beginning of life, there were only sensations and reactions by unicellular organisms. Sensing and reacting adequately started this way, messages were like irritating substances that caused the corresponding irritation. There were no "eyes" nor "ears", there were just the primordial of a perceiving process that, with evolution and with the development of nervous systems, would lead to world modelling, mind definition and, finally, subjectivity.

In a significant degree, the selectiveness that underlies the basic process of individuation and identification of meaningful patterns in the "environmental cauldron" by different cognitive entities, apparently requires no previous learning on the part of the cognitive system involved. It is as if all life forms naturally had "expectations" relative to the set of possible meaningful patterns to be found, at a particular stage of their development, in their environmental bubble, patterns corresponding to the potential satisfaction of existential requirements or configuring a threat to their existence.[5] In response to the agent's specific existential needs dictated by the status of its internal states, some environmental patterns become salient while all the other features, because they are life irrelevant, constitute "noise" that is ignored.

According to Thompson [50: 19], sense-making is threefold: (1) it is sensibility as openness to environment; (2) it is significance as positive or negative valence of environmental conditions relative to the norms of the living being; (3) it is the direction the living being adopts as response to significance and valence.

Apart from being embodied and embedded, cognition is also a situated process, i.e., it takes place in the particular context defined by the dynamics of both life form and that of its environment, a context where a dual existential narrative, a dual history gets defined.

In Dewey [10] terms, qualities emerge from organism-environment transactions and their proper locus is a situation, i.e., the overall experienced context or background within which properties, objects or relations are given.

On this topic, Varela writes:

> Ordinary life is necessarily one of situated agents […] situatedness means that a cognitive entity has by definition a perspective. This means that it isn't related to its environment "objectively" that is independent of the system's location, heading, attitudes and history. Instead, it relates to it in relation to the perspective established by the constantly emerging properties of the agent itself and in terms of the role such running redefinition plays in the system. [51: 30]

This context is in fact defined by the lived experience of the cognitive entity and by its present status, drives and "goals" in the particular circumstances that characterize the dynamics of the environment the system is embedded in. Bound by a semiosic relationship, both agent and its environment will "co-write" their respective existential narratives in a process where intelligence plays a fundamental role.[6,7]

---

[5] See footnote 4.

[6] Cf. Ferreira [18].

[7] For a detailed account and description cf. Ferreira and Caldas [16, 21].

2.3 The Specificity of Human Cognition

Though sharing the above mentioned three key existential constitutive conditions with all living entities, the human species distinguishes itself from all the others by having a set of specific attributes[8] namely: a conceptualising and symbolic capacity that allows for.

(i) the definition and reification of an external reality, (ii) the definition of a sense of alterity, of Otherness, (iii) the writing of an internal auto-narrative that permits the experiencer subject to recall past experiences, reflect on the present ones and project, predict and anticipate future events (iv) the sharing and transmitting of experience and accumulated knowledge, Ferreira [11, 13].

Cassirer [7] writes that whatever is alive has its own circle of action for which it is there and which is there "for" it—both as a wall that closes it off and as a viewpoint that it holds "open" for the world. However, according to him, only with humankind this life complex becomes a knowledge complex, i.e., an on-going individual and collective cumulative knowledge construction. Cassirer [7] defines the human being as "animal symbolicum" and proposes, on the basis of Uexkull's biology, the existence of a symbolic system, which falls between the "receptor" and "effector" systems that it shares with all of the other organisms. It is this symbolic system that allows signs to be assigned values, enhancing a three-part relationship between the "Sign-Using Self", "Constructed Reality" and the "Other Self". On the particular role of language, as a tool for the definition of a human-sized reality and for the sedimentation and sharing of knowledge, Cassirer [6] writes that naming singles out the particular aspects of the passing contents which never recur with any strict uniformity, providing it with a stable sign, and on the basis of this, an "artificial" unity that allows consciousness to raise itself to the sphere of objective thinking. He points out that we must break radically with the presupposition that what we call the visible reality of things is given and present at hand as a finished substratum prior to all formative activities of the mind, because it is not the reality of things which endures but only the form that reality assumes through us.

This constructed reality is not just a reflection that mirrors in our eyes. It is the image of the world as defined by human beings, a synchronic view shared by all the members of a community, at a given stage of their on-going developmental narrative, a reality determined by the individual human condition within a particular social and cultural setting. When we take a diachronic view and compare the still close reality of our grandparents with our own, we realise how these realities already differ significantly. And they differ because, the semiosic process underlying

---

[8] According to Ferreira [19] these human specific attributes, that in some cases exist in very incipient forms in other species, are (i) a conceptualising and symbolic capacity, (ii) a tool making capacity and (iii) the generative goal-oriented productive action—the transformative power of work. Cf: "On Human Condition", Proceedings of the 6th International Conference on Robot Ethics and Standards-ICRES 2021. https://www.clawar.org/icres2021/ and "On Human Condition: The Status of Work" (2021) in Maria Isabel Aldinhas Ferreira and Sarah Fletcher eds, The 21st Century Industrial Robot: When Tools Become Collaborators, Springer, ISCA series.

the agents' interaction with specific physical, economic, social and cultural environments, produces world views that are made of distinct constellations of individual entities. These constellations comprehend not only distinct physical objects, namely artefacts, but also different patterns of behaviour and social conventions, different lifestyles, distinct institutions, specific cultural routines, rituals and celebrations as well as differentiated architectural options.

These conventions are generated, defined and established inside the community's specific life contexts, i.e., in the life contexts that emerge within specific physical, economic, social and cultural frameworks.[9]

Varela refers that

> our lived world is so ready-at-hand that we are in no way deliberate about what it is and how we inhabit it. When we sit at the table to eat with a relative or a friend, the entire complex know-how of handling table utensils, the body postures and pauses in the conversation, are all present without deliberation [52: 328]

These conventions, that constitute semiosic matrices, are incorporated in the course of the individual's lifetime, within the multiple contexts of experience of their lived world—the initial home context, followed by all the others the individual comes across throughout their personal and social development. Simondon [46] calls these established conventions the pre-experiential background issued from the accumulated experience of precedent generations, a common background that only comes to life in each present individual appropriation, being in this way also consequently affected by the action of those presently making use of it.

Human reality, in the multiple social and cultural forms it has been assuming throughout historical times, is the semiosic outcome of the individual and collective interaction of intelligent self-conscious organisms with the multifaceted physical, social and cultural environments they construct and are embedded in.

2.4 *Representing Life Dynamics: The Need for a Non-mechanistic Approach*

Glennan [26] refers that models are representations that describe (in some degrees and respects) the mechanism that is responsible for a given phenomenon.

The mechanistic perspective on life dynamics has been present, since the seventeenth century, through Descartes' metaphor of the organism as clockwork. As technology developed, in the nineteenth century, the steam engine became the preferred metaphor, being replaced later by the computer—the so far "unchallenged candidate", Hoffmeyer [30].

But to reduce life dynamics to these metaphors leads to a very impoverished perspective that misses fundamental components of the life phenomena under

[9] Interpretation signifies here assigning a certain value-meaning to an environmental feature or pattern.

scrutiny, namely the essential semiosic process that binds the organism to its surrounding world and is responsible for the existential dialectics. In fact, semiosis is the life-sustaining process present in all forms of interaction and responsible for the organism's capacity to attune with its surrounding environment.

Living systems are not mechanisms, but as any mechanism they are made up of "purpose-structured components". However, opposed to the typical rigidity of man-made mechanisms, they are endowed with a plasticity that allows for their intelligent participation in the semiosic process grounding the dialectics life depends on, being this way not only capable of recognizing and responding to meaningful patterns but also of adapting and ultimately evolving in order to cope with prompts placed by the contextual settings they are embedded in. With the emergence of Sars-Covid in the human environment, we have been having the privilege to observe (though also enduring its effects) how a life form, a virus, has been able to evolve in our human environment, adjusting, adapting and mutating in order to optimise its replication capacity.

We would say that a model that intends to replicate a life process should always map the way a given system, a specific purposeful organisation, is structured and its extended internal and external dynamics, i.e., the type of interactions it develops in order to achieve its purposes and/or the purposes of the larger system/organisation it is part of.

We believe that in order to translate the dynamics present in the semiosic process responsible for the dialectics cognitive agent/environment, i.e., the dialectics that binds life forms with their lived worlds, one has to model how a particular corporeal architecture shapes in the general environment a particular meaningful sphere—its Umwelt—and the extent of its actions in this microcosm.

Acknowledging the specificity of human cognition within the universal framework defined by the three key existential constitutive conditions common to all life forms,[10] and acknowledging semiosis, i.e. meaning-assignment to specific environmental patterns by an intelligent entity, as the dynamo of cognition, Ferreira and Caldas [21] proposed the modelling of natural cognition, as depicted in Fig. 1.

According to this model[11]:

- $\boldsymbol{y}$ is a vector of dimension: (Ny $\times$ 1), which is assumed to represent all the potential information available for all living entities capable of evolving in a given environment.
- Since not all environmental features potentially present in that environment will be perceived/sensed by the cognitive agent, due to the specificity of its corporeal architecture, and since even the perceived/sensed features will be given different degrees of relevance at different times and within different contexts, according to the agent's internal states, the agent's "view" of its environment (Umwelt) was modelled through a (N**u** $\times$ 1) vector, $\boldsymbol{u}$.

[10] Cf. Ferreira [15].

[11] Cf. Keucheyan [17] on the concept of "besoin".

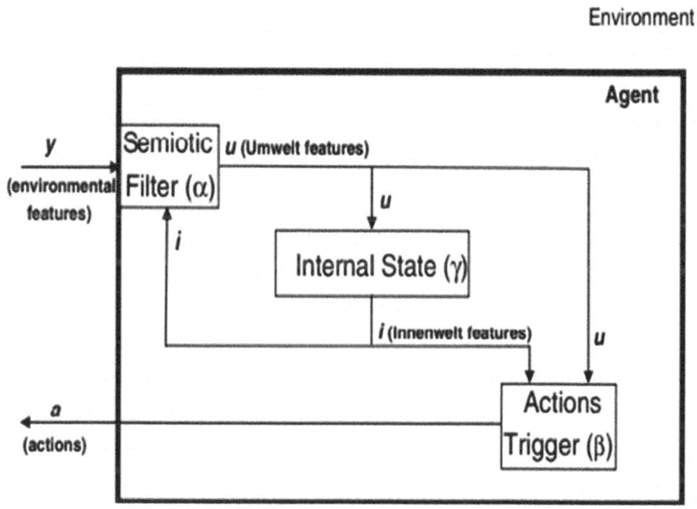

Fig. 1 Modelling natural cognition

- Vector $\boldsymbol{u}$ is created from the general—environmental—features vector, $\boldsymbol{y}$, through the application of a semiotic filter, **F**, whose characteristics are dependent on the agent's nature and internal states (Innenwelt), represented through a ($\mathbf{Ni} \times 1$) vector, $\boldsymbol{i}$. The perceived/sensed patterns modelled through the ($\mathrm{Nu} \times 1$) vector, u will have different relevance and consequently different salience depending on the urge for satisfaction of the needs determined by the agent's internal state(s).

  In the case of human beings the set of these needs is well beyond the mere satisfaction of basic essential physical needs, such as having food, shelter… or even psychological/emotional urges, such as belonging, loving and being loved… Being the result of a process of cultural incorporation and integration in a specific social and cultural environment, a huge set of "complementary" needs are civilizational/culturally determined and depend on the community's economic, social and cultural specificity at a particular stage of its economic and historical development.[12]

  The semiotic filter, F, that allows for the definition of vector u retains not only the species-specific sensible biological patterns but through symbolic encoding and thanks to human conceptualising-linguistic capacity, also the prevailing social and cultural patterns embodying the values taken as significant for a certain community.
- This means that when analysing human cognition, i.e., when identifying individual and/or collective human behaviour in particular frameworks, it has to be taken into account the specificity of the contextual variants involved and how they have been incorporated or not by the agent, that is to what extent individual and collective behaviour is affected and determined by social and cultural values/conventions.

[12] See footnote 11.

- Meaning assignment in the course of the essential semiosic process will trigger the agent's actions/behaviour(s) and cause the consequent transition to a new internal state.
- This new internal state whose duration will depend on the corporeal dynamics of the agent and on the degree of stability offered by its surrounding environment, will correspond generally to the satisfaction of the precedent basic need (for energy, for replication, for defence against a predator/threat...) or in the case of human beings in the satisfaction of multiple higher level urges as that of getting an education, having a proper job, being recognized by performance or achievement....
- As a consequence of the organism's requirements and/or of its interaction with the surrounding environment, new internal states will emerge afterwards, influencing both the agent's perceptions and its reaction towards new salient sensed environmental patterns.
- The vectors ***u*** (Umwelt) and ***i*** (Innenwelt) are, therefore, in a dialectic relationship that determines and triggers the agent's (re-)actions. We assume that there are **Na** possible actions that can be executed by the agent and collect the respective probabilities of execution in a vector, ***a***. These actions, when executed, will have an effect on the environment, allowing or not the satisfaction of the needs dictated by internal states and eventually providing a means for learning to occur.

Assuming semiosis as the dynamo of cognition is essential when attempting to model any form of hybrid or exclusively artificial cognition. This will render purely mechanical and frequently partial descriptions, just understood in terms of cause/effect relationships into comprehensive views of the functioning of the organic whole.

Acknowledging the extreme relevance of the phenomenon under analysis and its levels of complexity will contribute not only to a better understanding of all the variants involved in the different life phenomena, (either biological, social or cultural) their significance and interplay, but will also highlight how fundamental this understanding is in the construction of efficient artificial forms of cognition, artificial forms of intelligence, tuned up with the settings of the specific organisation they will be embedded into and the semiosic matrices of the society in which they will be deployed.

3 The Task of Replicating Intelligence

3.1 First Steps and the Way Forward

In 1956, at Dartmouth summer school, McCarthy and a group of researchers[13] coined the term Artificial Intelligence claiming that machines could simulate every aspect of learning or any other feature of intelligence.

[13] Cf. McCarthy [36].

Limiting the concept of intelligence to human intelligence and taking the human brain as the source of inspiration for AI modelling, McCarthy et al. identified as instances of intelligence the use of language, the formation of abstractions and concepts, the capacity for solving problems and for self-improvement. However, McCarthy's concept of intelligence, working in fixed-ruled environments, showed itself unable to cope with unpredictable, unstructured environments, which are in fact the ones living entities are embedded in and have to interact with.

A new concept of intelligence was consequently needed, one with enough plasticity to be able to respond and eventually adapt to the unstructured nature of the ever evolving surrounding environment, capable of reacting adequately to prompts as living organisms do. In our words, one capable of engaging in a semiosic relationship with the dynamic environmental context.

In the last decades, humankind has been experiencing the cumulative effects of a profound technological development, one that has introduced new tools, new channels and new forms of communication, that has been reshaping the human sphere of action and interaction; one that has hybridised human reality by merging the natural and the artificial, the real and the virtual and that has probably started a new civilizational stage [18].

This technological (r)evolution, that has been empowering human cognition by accelerating immensely its intrinsic semiosic processes, has in the last decades been affecting, altering the very nature of the typical human Umwelt by incorporating the digital in the surrounding physical environment and by creating new possible forms of agency and consequently new lived worlds.

According to Ferreira [17, 19], the deployment of artificial cognition in the twenty-first century, has introduced a fundamental twofold ontological shift: (i) tools have lost their purely instrumental status and have suddenly become potentially autonomous entities capable of a form of shared agency when co-acting with humans and even capable of decision making in multiple circumstances (ii) environments, on the other hand, have acquired a hybrid nature, where the analogic and the digital merge and where the physical and the virtual converge, where natural intelligence and autonomous artificial systems cohabit " in a fusion that blurs the lines between the physical, digital and biological".[14]

As Klaus Schwab pointed out on the 14th January 2016[15]:

> We stand on the brink of a technological revolution that will fundamentally alter the way we live, work, and relate to one another. In its scale, scope, and complexity, the transformation will be unlike anything humankind has experienced before.

[14] https://www.weforum.org/agenda/2016/01/the-fourth-industrial-revolution-what-it-means-and-how-to-respond/.

[15] ibidem.

3.2 The Natural, the Artificial and Their Hybrid Forms

Addressing the impact of the incorporation of autonomous artificial systems in human reality, [17] identifies the main types/forms of semiosic interactions, i.e., the potential distinct forms of cognition present in this stage of technological development:

(a) The Natural Forms of Cognition (Neither the Agent nor its surrounding Environment have been object of any transformation or enhancement through the incorporation of artificial systems). These comprehend:

- The typical dyadic forms involving a natural system and its physical environment as the one that characterises the life dynamics at micro or macro level of entities such as cells, bacteria, viruses, animals, plants…
- The more sophisticated form of natural cognition that characterises the being in the world of humankind and that being symbolically encoded, has a triadic nature.
- The interactions that take place between life forms (human beings) and physical environments that are not their typical ones, as the interactions occurring in outer space, in different planets, or just in a different milieu, e.g. underwater,

(b) The Hybrid Forms (either the agent or its environment have a digital system incorporated in its architecture), namely:

- The forms involving natural systems—human beings—and digital interfaces existing in the analogue world, in typical human life contexts.
- Those instances involving human beings interacting with virtual environments or augmented reality scenarios … where displacement from the subject's actual spatio/temporal framework occurs, as those induced by electronic devices operating on the external perception organs or through induction in the neural system.
- The forms involving human beings with enhanced capacities and the typical natural physical environment, as in the case of bionic components.
- The forms involving an embodied artificial intelligent system evolving and acting in a humanised physical and social world as in the case of robotic systems co-acting with human beings at the work place or performing social routines in public spaces or at home.

(c) The Exclusive Artificial Forms of Cognition: Those forms in which both agent and its environment are digital entities.

- The forms actualised by algorithms that evolve in the context of data related to specific domains: economy, health, business…, analysing recurrent patterns, identifying trends, diagnosing, predicting behaviours or outcomes and making subsequent proposals or decisions.

- Those that take place when objects that are part of human daily life are capable of digitally interacting by exchanging information among themselves and also with the user as in the case of the Internet of Things (IoT).
- Those forms of interaction that are meant to replicate life through the constitution of artificial environments and artificial agents populating these environments and interacting with each others, as in the case of Artificial Life.[16]

In the next section, we will briefly address the specific semiosic relationship that emerges from the interaction of a digital agent (an algorithm) and its Umwelt—the digital environment in which it evolves. Because AI/ML comprehends a plethora of different applications suited for different domains, we will only consider the particular case of digital marketing.

3.3 *When Algorithms Couple with Data: The Digital Umwelt*

Data and statistics have always provided the essential basis to understand the ontogeny of systems and their evolution within organisms or organisations. A synchronic perspective takes a snapshot to the status of a system, momentarily freezing its dynamics, in order to be able to identify and quantify the patterns emerging in a given informational environment and the interdependencies established between them. The diachronic perspective monitors the recurrent or punctual character of those patterns and how they have been interacting and constraining each other, predicting the emergence of potential new trends or outcomes within the overall processual dynamics. Both perspectives are essential to the management and decision making within organisations and they have been around in analogue form, for decades, particularly in marketing, economics and finances.

The first wave of digitization started a progressively comprehensive replacement of analogue information processing thanks to the digital capture and storage of data. This process grew exponentially with the increasing computational power of machines accompanied by their massive storing capacities that allowed information, coming from multiple digital sources, to be collected and to be stored as data that could be filtered, categorised and processed at incredible speed, providing live access to an ever-evolving digital environment, available at any time and from everywhere.

A dense ecosystem of technologies is, at the moment, responsible for collecting about 2.5 quintillion bytes of raw data created each day. Providing analytical insights on such a huge amount of data, in real time, requires strong computational processing power and specific tools. Machine Learning is, presently, the key technology for the creation of predictive models capable of assisting in the solution of complex problems across nearly every domain of human existence, such as governance, business

[16] Cf. Scott et al. [42].

and management, healthcare, education, criminal justice. These contexts—specific digital environments in fact constitute digital Umwelten where dedicated algorithms evolve performing, in specific settings, the particular functions they were designed for.

Data is usually referred as the fuel of economy or as the new oil in the sense that business, enterprises and financial markets will be able to grow exponentially by better managing their own strategies and resources and by being able to anticipate/predict consumers' trends or definitely influence consumers' habits in a much more "efficient" way than ever before. However, Big Data, in the sense of raw data, is presently useless unless properly handled, i.e. it has to be previously categorised and tagged by human beings so that algorithms can be trained being subsequently capable of evolving in a specific digital environment where they will be able to attain their design objectives. According to Guest Post: A Guide to Leveraging Active Learning for Data Labelling (The Sequence, Dec, 2021), nowadays teams spend about 80% of their time building and maintaining training data infrastructure, including cleansing, transforming, and labelling data in preparation for model training. Efficient algorithmic functioning and optimization depends not only on the intelligent design of algorithmic architectures by human beings but also on time consuming and exhausting categorising work by human intelligence in order to constitute suitable training datasets.

The optimization of this form of artificial cognition in fact depends on human following a set of procedures all along the processual pipeline, as it can be exemplified by taking the case of in business management but that can easily be adapted to any other field:

1. An initial accurate assessment of the organisation's functional structure and the identification of the particular sectors AIS performing could represent an added value.
2. The previous assessment of the impacts on the organisation's functional structure by the deployment of such tools in order to take the necessary measures to nullify potential negative consequences.
3. The identification and quantification of the main benefits this deployment, in (a) specific sector(s), could bring about.
4. The design of algorithmic architectures embodying the organisation's prevalent values and goals with consideration to the semiosic matrices that characterise the regional or universal social and cultural context.
5. The constitution and permanent updating of representative datasets that privilege quality over quantity in order to get the best possible model training.
6. Data observability. According to Gavish [24] this observability concerns the following topics: **Freshness**: How up-to-date are your data tables? **Distribution**: Do your data values fall within an accepted range? **Volume**: Are your data tables complete? **Schema**: Who made what changes to your data, and when? **Lineage**: The full picture of your data landscape, including upstream sources, downstream ingestors, and who interacts with your data at which stages.

Digital marketing operates, via multiple digital media, aiming to identify the needs, preferences and interests of individual end users/potential consumers. By tracking their digital moves throughout the web it creates individual profiles that will feed the system leveraging a personalised approach and individualised interaction on the part of the interface based on those features that give substance to a specific profile. Simultaneously, this massive individual tracking will allow for the identification of specific trends and drives in a population according to demographics, location cultural variability.

But digital marketing not only aims to identify present patterns of behaviour, wishing to be able to predict new future ones, but has also the goal of acting on that behaviour by influencing and guiding end users/consumers's choices in specific ways.

We will finish our chapter by reflecting on how human intelligence and AIS co-act, in the digital marketing industry. This domain is probably the richest evidence provider in what concerns the way AIS are incorporated in human cognition and semiosic processes. That incorporation determines and affects two different levels of semiosic relationship: (i) the one that is established by the potential consumer and the digital interface they are interacting with, which acts like a virtual Umwelt (ii) and the one that is defined by the algorithms interaction and with their digital environment within this interface.

The following passage by Schmarzo [41] seems very illustrative of these phenomena:

> In the digital marketing industry, companies have a fraction of a second to determine a visitor's intent when they visit a site – that is, what is the visitor trying to accomplish – in order to guide decisions about
>
> 1) what to bid on that visitor (real-time bidding for the right to show them an ad),
> 2) what ad to show them
> 3) where to place that ad on the site.
>
> To determine each visitor's intent, digital marketing companies accumulate a deep history of each visitors' interaction activities including what sites they visited, what sites they came from, what sites they went to next, how long they spent on each site, what ads they clicked on, what ads they did not click on, and their search requests. They analyse the visitor's interaction history to determine and codify (propensity score) their intentions (e.g., most likely interested in a car, vacation, sporting event, etc.) that drives the real-time ad placement decisions.

Schmarzo [41] points out that to create an intelligent user experience requires leveraging AI/ML to analyse a deep history of the user's interactions to determine the user's intentions, and then coupling those intentions with current trends, patterns, and relationships to match those intentions with a deep understanding of the available content to recommend the most relevant action.

According to Schmarzo (ibidem) creating an Intelligent user experience requires

- **Intent Determination**. What is the User trying to accomplish and the associated decisions, pains, and gains?

- **Value Definition.** What is the "Value" of accomplishing that task (their intentions) to the User and what are the KPIs and metrics against which the User will measure the effectiveness?
- **Value Creation.** How is the organisation capturing and codifying User Intent and matching that intent to the most appropriate content and actions?

We will not address here the so much debated ethical issues emerging from this systematic "electronic tracking" or should I say "stalking" and not even how this permanent channelling of the user to particular objects, entities or events in fact heavily constrains their chances of looking for, finding or discovering something else per themselves, something that may be totally out of their common habits, trends or favourites, regrettably limiting their horizons, their intelligence, ultimately menacing to manipulate in a more powerful way than ever the "free" construction of a vision of the world.

So contrary to Schmarzo we do not agree that we are creating an intelligent user experience. We are in fact creating a very dumb and limited user experience, limiting their digital Umwelt to what they already are or already think they "know" by using our limited intelligent tools.

4 Conclusion

Questioning what is at stake when we talk of intelligence, we tried to highlight that intelligence is an attribute of all living systems, an attribute that allows them to evolve in their respective environments, to respond to the prompts those environments pose, to adapt to new circumstances, to subsist and replicate.

To fully understand this dynamics we believe one has to be aware that cognition is always an embodied, embedded and situated phenomenon that in the case of human cognition assumes an upper level of semiosic complexity granted by human conceptualising/symbolic capacity and by the generative capacity to act and transform the world by creating tools whose degree of sophistication has been increasing throughout historical time.

This means that human reality is a multilayered construct and that our vision of the world is determined by human biology and ontogeny in interaction with the specific physical, economic, social and cultural context individuals are immersed in, in other words the human vision of the world is grounded on biological and culture—determined semiosic matrices whose substance differs not only according to a diachronic perspective, but also synchronically. Even in a globalised world we find distinct cultures, distinct lifestyles, distinct values, distinct visions of the world. It is the dynamics of this ever-evolving cultural variety, this richness, that guarantees the exponential growth of human intelligence, the accumulation of knowledge and the capacity to develop and subsist.

Artificial intelligent systems in a way also have their own specific Umwelten—their meaningful worlds—these are nothing but a reflection of the semiosic matrices that constitute human physical and cultural reality. But opposed to life forms, artificial systems do not exhibit yet real intelligence, not even when compared with the most basic living entities, as they lack the capacity to autonomously respond to the dynamics of an ever evolving environment engaging with it in a truly dialectic relationship, in a semiosic relationship. And this is the requisite for being designated [intelligent]. In this sense, we completely agree with CNCDH (2022) when the report points out that there is an excess of anthropomorphisation in many terms relating to new technologies, namely in the terms [artificial intelligence], [deep learning] or [neural networks]. These terms inherently human are misleading relative to the nature of these artificial systems that are nothing more than computations, mathematical models and they are also misleading in what concerns the real performance capacities of these systems, namely ML systems. Still, according to the CNCDH report (ibd.,), because of its psychological impact, raising unfunded expectations or generating fears, this type of concepts should be avoided and replaced by neutral ones. As suggested artificial intelligent systems could be formulated in French as [SAAD]], i.e., [Systèmes algorithmics d'aide à la décision].

AIS are sophisticated tools and, as all the other tools human beings have created throughout historical times, extensions and empowerment of the human body. In this particular case, extensions of the way human beings view, construct and act in this world.

Aiming to be extensions of human intelligence, they are, in a way, partly embodied as they reflect and feed on human perception. In the same way, the substance of the Umwelt they are embedded in is also a reflection of a human world with its entities, its codes, its values.[17]

As Carpo [4: 81] refers:

> […] just like the Industrial Revolution created prosthetic extensions that multiplied the strength of our natural bodies, the digital revolution is now creating prosthetic extensions that multiply the strength of our natural intelligence […]

Together we will, hopefully, create a better world.

[17] Cf. Varela [51].

References

1. Barbieri M (ed) (2007) Introduction to biosemiotics: the new biological synthesis. Springer, Berlin
2. Barbieri M (2009) A short history of biosemiotics. Biosemiotics 2:221–245. https://doi.org/10.1007/s12304-009-9042-8
3. Barthes R (1973) Elements de semiology. Translated by Annette Lavers and Colin Smith. Hill and Wang, New York
4. Carpo M (2017) The second digital turn: design beyond intelligence. The MIT Press, Cambridge, MA
5. Carroll L (2000 [1871]) Alice's adventure's in Wonderland and through the looking-glass. Signet Classic. New American Library. Penguin Putnam Inc., New York, USA
6. Cassirer E (1985) The philosophy of symbolic forms. Vol 3. The phenomenology of knowledge. Yale University Press, New York
7. Cassirer E (1996) The philosophy of symbolic forms. Vol. 4. The Metaphysics of symbolic forms. Yale University Press, New York
8. Commission Nationale Consultative des Droits de L'Homme AVIS RELATIF À L'IMPACT DE L'INTELLIGENCE ARTIFICIELLE SUR LES DROITS FONDAMENTAUX, Avril 2020. https://media-exp1.licdn.com/dms/document/C561FAQE2RLaDKvZIcA/feedshare-document-pdf-analyzed/0/1649439818618?e=2147483647&v=beta&t=JTZxVV834zxVHJvLIIHXrsQ-2_939ui1pD9WWtBOVkE
9. Damásio A (2017) The strange order of things: life, feeling, and the making of cultures. Vintage
10. Dewey J (1958 [1925]) Experience and nature. Dover, New York
11. Ferreira MIA (2007) On meaning: the phenomenon of individuation and the definition of a world view. Ph.D. thesis. University of Lisbon, Lisbon
12. Ferreira MIA (2010) On meaning: a biosemiotic approach. Biosemiotics 3(1):107–130. https://doi.org/10.1007/s12304-009-9068-y
13. Ferreira MIA (2011) On meaning: individuation and identity. The definition of a world view. Cambridge Scholars Publishing, Cambridge
14. Ferreira MIA (2012) Modelling artificial cognition in biosemiotic terms. Biosemiotics 6:245–252. https://doi.org/10.1007/s12304-012-9159-z
15. Ferreira MIA (2013) Typical cyclical behavioural patterns: the case of routines, rituals and celebrations. Biosemiotics 7:63–72. https://doi.org/10.1007/s12304-012-9159-z
16. Ferreira MIA (2015) Semiosis: the dialectics of cognition. In: Trifonas P (ed) International handbook of semiotics. Springer, Netherlands. ISBN-13: 978-9401794039. https://doi.org/10.1007/978-94-017-9404-6
17. Ferreira MIA (2018) Cognitive architectures: the dialectics agent/environment. In: Ferreira MIA, Sequeira J, Ventura R (eds) Cognitive architectures, intelligent systems, control and automation: science and engineering, vol 94. Springer. https://doi.org/10.1007/978-3-319-97550-4_14
18. Ferreira MIA (2020) How smart is your city? In: Ferreira MIA (ed) Technological innovation, ethics and inclusiveness. Springer, Cham. ISBN: 978-3-030-56925-9. https://doi.org/10.1007/978-3-030-56926-6
19. Ferreira MIA (2021) On human condition. In: Proceedings of the 6th international conference on robot ethics and standards—ICRES 2021. https://www.clawar.org/icres2021/
20. Ferreira MIA (2022) On human condition: the status of work. In: Ferreira MIA et al (eds) The 21st century industrial robot: when tools become collaborators. Intelligent systems, control and automation: science and engineering series. Springer
21. Ferreira MIA, Caldas MG (2012) Modelling artificial cognition in biosemiotic terms. Biosemiotics (August 2013) 6(2):245–252. Springer. ISSN 1875-1342. https://doi.org/10.1007/s12304-012-9159-z
22. Fleischemann RW Jr (1996) Viral genetics. In: Baron S (ed) Medical microbiology, 4th edn, Chapter 43. University of Texas, Medical Branch at Gavelstone

23. Fox E (2005) Ecosystems, organisms and machines. BioScience 55(12):1069
24. Gavish L (2021) What is data observability and why it matters. In thenewsack.io https://thenewstack.io/what-is-data-observability-and-why-does-it-matter/ (April 2021)
25. Gibson J (1979) The ecological approach to visual perception. Houghton Mifflin, Boston
26. Glennan S (2017) The new mechanical philosophy. Oxford University Press
27. Gottlieb G (1971) Ontogenesis of sensory function in birds and mammals. In: Tobach E, Aronson L, Shaw E (eds) The biopsychology of development, vol 81. Academic Press. New York
28. Haenlein M, Kaplan A (2019) A brief history of artificial intelligence: On the past, present, and future of artificial intelligence. Calif Manag Rev 61(4):5–14. https://doi.org/10.1177/0008125619864925
29. Hoffmeyer J (2008) Biosemiotics: an examination into the signs of life and the life of signs. University of Scranton Press, Scranton/London
30. Hoffmeyer J (2012) The natural history of intentionality, a biosemiotic approach. In: Schilhab T et al (eds) The symbolic species evolved, biosemiotics, vol 6, 6 C. Springer Science+Business Media B.V. https://doi.org/10.1007/978-94-007-2336-8
31. Lecanuet JP, Granier-Deferre C, Busnel MC (1995) Human fetal auditory perception. In: Lecanuet JP, Fifer WP, Krasnegor N, Smotherman WP (eds) Fetal development—a psychobiological perspective. Lawrence Erlbaum Associates, Publishers, New Jersey
32. Li D, Du Y (2007) Artificial intelligence with uncertainty, 1st edn. Chapman and Hall/CRC. https://doi.org/10.1201/9781584889991
33. Lickliter R (1995) Embryonic sensory experience and intersensory development in precocial birds. In: Lecanuet JP, Fifer WP, Krasnegor N, Smotherman WP (eds) Fetal development–a psychobiological perspective. Lawrence Erlbaum Associates, Publishers, New Jersey
34. Littman ML, Ajunwa I, Berger G, Boutilier C, Currie M, Doshi-Velez F, Hadfield G, Horowitz MC, Isbell C, Kitano H, Levy K, Lyons T, Mitchell M, Shah J, Sloman S, Vallor S, Walsh T (2021) Gathering strength, gathering storms: the one hundred year study on artificial intelligence. (AI100) 2021 study panel report. Stanford University, Stanford, CA, September 2021. Doc. http://ai100.stanford.edu/2021-report. Accessed 16 Sept 2021
35. MacKenzie M (2016) J Philos Stud 31:21–36
36. McCarthy M (2006) A proposal for the Darmouth Summert Research Project. https://ojs.aaai.org//index.php/aimagazine/article/view/1904
37. Nagel T (1974) What is it like to be a bat? Philoso Rev 83(4):435–450
38. OECD (2019) Artificial intelligence in society. OECD Publishing, Paris. https://doi.org/10.1787/eedfee77-en
39. Peirce CS (1932) The collected papers of Charles Sanders Peirce, 8 vols. Harvard University Press, Cambridge, MA
40. Schaal B, Orgeur P, Rognon C (1995) Odor sensing in human fetus: anatomical, functional, and chemoecological bases. In: Lecanuet JP, Fifer WP, Krasnegor N, Smotherman WP (eds) Fetal development—a psychobiological perspective. Lawrence Erlbaum Associates, Publishers, New Jersey
41. Schmarzo B https://www.datasciencecentral.com/profiles/blogs/the-power-of-determining-user-intent (6 November 2021)
42. Scott R, MacPherson B, Gras R (2018) EcoSim, an enhanced artificial ecosystem: addressing deeper behavioral, ecological, and evolutionary questions. In: Ferreira MIA, Sequeira J, Ventura R (eds) Cognitive architectures, intelligent systems, control and automation: science and engineering, vol 94. Springer. https://doi.org/10.1007/978-3-319-97550-4_14
43. Sebeok TA (1976). Studies in semiotics. Contributions to the doctrine of signs. Indiana University Press, Bloomington, pp 1–45
44. Sebeok TA (1991) Semiotics in the United States. Indiana University Press, Bloomington
45. Sequence. https://thesequence.substack.com/p/labelbox?token=eyJ1c2VyX2lkIjoxODQ5OTk1MywicG9zdF9pZCI6NDQ3OTQ5ODcsIl8iOiJIUnhyQyIsImlhdCI6MTYzODM2NjEyMiwiZXhwIjoxNjM4MzY5NzIyLCJpc3MiOiJwdWItNTQzMDkiLCJzdWIiOiJwb3N0LXJlYWN0aW9uIn0.2T5n22BfDWuAC7U3CL8YxYl4UiQ_Oaeo_bEtq_CfmCE

46. Simondon G (1964) L'Individu et sa genèse physico-biologique. PUF, Paris
47. Smotherman WP, Robinson SR (1988) The uterus as environment: the ecology of fetal behavior. In: Blass EM (ed) Handbook of behavioral neurobiology, vol. 9, Developmental psychobiology and behavioral ecology. Plenum, New York
48. Smotherman WP, Robinson SR (1995) Developmental trajectories. In: Lecanuet JP, Fifer WP, Krasnegor N, Smotherman WP (eds) Fetal development–a psychobiological perspective. Lawrence Erlbaum Associates, Publishers. New Jersey
49. Stahl B (2020) Artificial intelligence for a better future: an ecosystem perspective on AI ethics and digital emerging technologies. Springer Briefs in Research and Innovation. Springer International Publishing
50. Thompson E (2011) Living ways of sense making. Philos Today 55(Supplement):114–123
51. Varela F (1992) Autopoiesis and a biology of intentionality. In: Proceedings from the Dublin workshop on autopoiesis and perception, essay 1. www.eeng.deu.ie/pub/autonomy/bmem9401
52. Varela FJ, Thompson E, Rosch E (2017, 1991) The embodied mind: cognitive science and human experience. MIT Press
53. von Uexküll J (1973 [1928]) Theoretische Biologie. Frankfurt am Main, Suhrkamp
54. von Uexkull J (1933) A theory of meaning. In a foray in the world of animals and humans. In: Wolfe C (ed) Posthumanities, vol 12. Minnesota Press 2010.
55. White Paper on Artificial Intelligence Feb 2019, European Commission

In Machines We Trust?

Maria Isabel Aldinhas Ferreira

Abstract The semantic substance that gives body to the concept of [Trust], and its slight variation according to the distinct contexts of experience and/or according to the different entities involved, is here briefly addressed in order to identify what is really at stake when we define [Trust] as a core criterion for accepting the development and deployment of artificial intelligent systems (either embodied or non embodied) in different domains of human life.

Highlighting the deep biological roots the concept stems from, the paper posits the existence of a preconceptual primitive common to most living organisms. Human conceptualising capacity incorporates this primitive and through a process that is socially and culturally determined shapes its rich meaning-nuanced character.

Claiming that the concept is primarily inherently associated with an implicit or explicit risk assessment by the potentially affected entity, the paper identifies two fundamental cognitive dimensions involved in that assessment: an intuitive and a rational. The rational risk assessment bases itself on the individual's direct experience or on the empirical evidence provided by the experience of others. In cases such as those of adopting a scientific or technological innovation, and lacking the competence to ground their judgement, the end-user's assessment will rely heavily on a certification issued by organisms, publicly acknowledged as competent, that declare the risk free character of a product or procedure.

It is in this context that benchmarking, standardisation and certification assume a fundamental role across the processual pipeline. The permanent and systematic monitoring of how all these new technologies evolve in their different contexts of use, updating predefined sets of standards will guarantee their quality, efficiency, reliability, integrity and safety.

The paper concludes by advocating the urge of making standards in general, and the standardisation of AIS in particular, available as open-source for the sake of a universal harmonious and beneficial technological development.

M. I. A. Ferreira (✉)
Centre of Philosophy of the University of Lisbon, Lisbon, Portugal
e-mail: isabelferreira@letras.ulisboa.pt

Institute of Systems and Robotics/Instituto Superior Técnico, University of Lisbon, Lisbon, Portugal

© The Author(s), under exclusive license to Springer Nature Switzerland AG 2022
M. I. A. Ferreira and M. O. Tokhi (eds.), *Towards Trustworthy Artificial Intelligent Systems*, Intelligent Systems, Control and Automation: Science and Engineering 102,
https://doi.org/10.1007/978-3-031-09823-9_2

1 Introduction

When starting to write these lines immediately came to my mind the segment "In God We Trust" printed in the US metal and paper currency and acknowledged since 1956 as the US national motto. Despite the continued arguments against the inappropriate association of religious beliefs with political institutions, the motto is still acknowledged, by a vast majority of the American people, as substantiating their fundamental values as a society.[1]

Imprinted on the object that is present in all their trivial daily transactions and that guarantees individual subsistence, this short but powerful statement assumes itself with the force of a collective belief–Faith-on the role of transcendence protecting and favouring human beings and their enterprise.

Technological development and innovation are frequently seen, especially by those dedicated to it, as the way for human beings to achieve well-being and happiness, a fundamental conviction on the benefits that technology can bring to humanity, nearly a faith in the way machines, namely intelligent machines, will foster a better world.

This fact led me to map that collective faith commitment, onto a totally different sphere questioning, this way, the nature and limits of the concept of [Trust] when applied to the so-called intelligent machines.

2 The Biological Roots of Trust

Life forms, contrary to lifeless systems, evolve in the environment they are embedded in by establishing their own interactions and dynamics, actively searching for the resources required to subsist, and simultaneously trying to avoid or overcome the potential threats to their existence.

That dynamics demands on the part of the embedded cognitive entity a permanent awareness to its surrounding world in order to spot and identify the emergence of patterns that configure the potential satisfaction of the requirements dictated by its physiology and simultaneously an assessment of the potential risks to its integrity posed by the presence of eventual nocive entities.

This awareness guarantees the timely triggering of the adequate behaviours by the organism in order to satisfy existential requirements such as those corresponding to the parameters energy or replication and the adoption of (re)actions—avoidance, camouflage, attack—meant to guarantee the safety of its physical integrity relatively to what is perceived as an external threat.

The dual activity of both scanning for the patterns that guarantee existence and replicability, as well as that of identifying and assessing the potential risk- carriers is present in the most basic to the most complex forms of cognition. Le Roux et al. point out in Bacterial Danger Sensing [29], bacteria face constant threats to their

[1] Cf., https://en.wikipedia.org/wiki/In_God_We_Trust.

survival due to extreme temperature, nutrient limitation, radiation and, last but not the least, the antagonism of other organisms. As a response to biotic threats that derive from bacterial competitors, they have developed, along their evolutionary narrative, generalised cellular programs, danger sensing pathways for counteracting them. Le Roux et al. (ibid.) identify the following elements of this program:

1. Bacteria can sense and respond to exogenous molecules associated with danger.
2. Danger signals can be damage-associated molecules derived from kin cells.
3. Various pathways transduce danger signals to launch protective cellular responses.
4. The response to threat may comprehend offensive and/or defensive behaviours.

The threshold of acceptance of co-existing with other organisms can be redefined in particular circumstances as it happens with animals that have been subjected to a process of domestication. In fact human beings throughout their developmental narrative have brought a wide range of species into domestication to use as livestock, working animals, household pets, Zeder [43] leading to a lowering of their defensive threshold in what relates to people. The influence of human behaviour on domesticated animals, namely household pets, has allowed species to learn to co-exist, sometimes leading to the formation of an interspecies affectionate relationship. This fact shows how the environment, as it always happens, has played a role all along the "evolutionary timeline" redefining the defensive threshold levels so that some species could incorporate the possibility of human presence as nearly risk-free, "trusting" them and allowing interaction to happen.

3 Trusting: The Glue of Interpersonal Relationships

[Trust] is a human concept[2] that stems from strong innate biological roots. Grounded on an innate system of self-preservation we share with other living entities, it has evolved to guarantee the establishing of a cooperative mood and the definition of temporary or enduring interpersonal relationships[3] necessary to the success of an inherently social species, necessary to individual and collective survival.[4] In the words of Kenneth [3] trust is an important lubricant of social systems.

According to Riedel and Javor [34] the emergence of trust behaviours results from (i) the close interaction between an innate system defining the loop-genes, hormones and brain functioning—with (ii) the environmental influences of nurture (deriving from a process of socialisation that takes place in particular contexts) cf. Bjornskov [5], Welch et al. [41], Johnson [28]. As Riedel and Javor (ibid) point out empirical

[2] For a comprehensive view on the different philosophical approaches to the concept cf. https://plato.stanford.edu/entries/trust/.

[3] According to Engelman and Herrman (2016) evidence indicates that human friendships have evolved especially robust forms of trust that are relatively immune to the contingencies of a volatile and ever-changing environment.

[4] Steven Pinker (2005) refers that there is evidence that trust has a strong evolutionary value.

evidence substantiates the notion that "trust behaviours" are the result of the complex interplay between both biological and environmental factors.

A person who meets a stranger for the first time intuitively reacts to their presence on the basis of stimuli such as appearance and facial expression, Righetti and Finkenauer [35], Todorov et al. [38]. As it happens in nature, the intuitive assessment performed is mainly an unconscious response to the way the Other is perceived and to the potential risk they may pose to one's integrity, or to one's goals in the short or long term. Other pre-epistemic judgements[5] may rest on context social/cultural defined parameters that come to define the trustworthy character of a stranger.

A global set of assumptions, based on what we could call a quick snapshot and formed largely on an intuitive basis, will determine the consequent trust behaviour-approach or distrust behaviour-avoidance.

As Riedel and Javor (ibid) refer, all components of a trust situation (from perceptions of stimuli such as the Other's facial expression to actual behaviour) are based on biological factors, namely the interplay of genes, hormones, and brain processes.

Damasio [11] supports the importance of automatic and unconscious information processing in trust situations and Zak [42] refers that trust is not a calculative activity, but a visceral sense that one has that a person can be trusted or not. Riedel and Javor (ibd.) conclude that a considerable proportion of neural processing in trust situations seems to be associated with automatic, emotional, and unconscious rather than deliberate, conscious information processing.

Though it can exhibit a semantically nuanced character depending on the language/culture it is encoded in, the concept of [Trust][6] is horizontal, i.e., it is transversal to different human communities throughout times, playing a fundamental role in the establishing of interpersonal, intersubjective relationships and cooperative ties essential to individual and collective development.

Social scientists have been for long trying to understand the mechanisms of human social interaction and cooperation. Their main analytical tool, Game Theory,[7] has been extensively used by disciplines such as economics, psychology, sociology and political science, focusing on the external determinants of human trust behaviour, namely the social and institutional environments. However, as Kosfeld (2007) points out, the internal biological mechanisms underlying and regulating individual trust-decision-making have remained more or less as a 'black box'. Acknowledging that, though being a fundamental component of human social interaction and interplay, trust is very difficult to quantify, Berg et al. [4] created the Trust Game to measure both trust and trustworthiness in an economic exchange.

[5] Those that don't result from actual personal or from reported experiences.

[6] Independently of its lec.

[7] Game theory has its roots in an article by the Hungarian mathematician John von Neumann, who subsequently published his book on the *Theory of Games and Economic Behavior* with Oskar Morgenstern in 1944. Game theory models a social situation as a strategic game. Such a game consists formally of three elements: the players who interact with each other; a set of available actions for each player; and a so-called pay-off function for each player. The possible outcomes that an individual player can achieve depend not only on the player's own behaviour but also on the other players with whom he or she interacts.

The development of technologies such as functional magnetic resonance or transcranial magnetic stimulation in the last decades combined with methods from both the social sciences and neuroscience has generated a very productive interdisciplinary field capable of trying to envisage how the human brain generates trust decisions. Within this framework, the trust game has been used successfully to study the neurobiology of human social behaviour.

The London Science Museum developed, in 2010, a version of the Trust Game as an educational tool so that visitors playing it could become aware of the role that neurobiology and also environment (social/cultural context) have in trust decision making and of the complex brain processes involved.[8]

In this game the interaction with the player starts with the statement:

> Trust is an important part of our everyday interaction with other people, followed by the question:
>
> > Who would you trust with your money?

A "trust situation" is then presented:

To loan a hundred pounds to **one** individual (from a set of possible four)

The game is played in four rounds. In each round the player has to choose the person they will be loaning the money to. At the end of each round the player is told the result of their option and gets a brief reflection on the internal (neurobiological) and external (environmental) factors involved when taking that particular decision.

In the first round the player is presented with the photographs of the four possible individuals to loan the money, together with their names: Claire, Winston, Jerome and Marge.

In the second round the same set of options is presented but now accompanied by each one's age: Claire-21; Winston: 36; Jerome: 40; Marge: 63.

In the third round information concerning their jobs is added: Claire-student; Winston-doctor; Jerome-artist and Marge-sweet shop owner.

Finally in the 4th round the player gets information about the purpose of the loan: Claire wants to start a charity for sick donkeys; Winston needs money to fund medical research into allergies; Jerome needs money to set up an art school; Marge needs the money to set up a new catering business.

As one can observe from this brief description, the choice in the first round is based on the candidates' appearance—their facial features. It is pre-epistemic as it stands on purely instinctive terms that warn of the potential risk of loaning money to each of those one is facing. In the second round, the choices are already made rationally and probably also culturally determined—some cultures may view older people as potentially more responsible and consequently more prone to pay back, others may privilege younger age as representing the fittest to satisfy the loan. The choices based on profession and purpose are also very rational though heavily context determined.

[8] The author played this game when visiting the museum in 2019.
What seems to be a video taken by another visitor when playing the game is presently available in the internet at https://www.youtube.com/watch?v=kzBzi8LNk34.

People having certain jobs, e.g., doctors, may be considered as more trustworthy than others while knowledge of the possible average income associated with each profession can also be determining for the choice. Finally trust—decision-making according to purpose can be determined on the grounds of rationality and/or by affective/emotional motives. In fact, while on one hand the investment on a certain project may sound as having more probabilities of being financially successful, being consequently more appealing for most trustors, the beneficial role of the project for society/humanity may surpass an exclusively determined investment option.

The Trust Game described here illustrates well how nature (the neurobiology responsible for human cognition) and the environment (the social and cultural contexts) interact with each other determining human decisions in what comes to identifying someone as trustworthy or rejecting others as distrustful. Trust and Distrust are in fact two sides of the same "coin", you either trust or do not trust someone, there is no middle point in this assessment. These options are part of an essential process of evaluation, always present in human interaction throughout life.

To conclude this section, we assume that the concept of [Trust].

(i) develops from a pre-conceptual primitive responsible for the defensive behaviour inherent to most life forms.[9]

(ii) has evolved in human beings as a consequence of the need to establish temporary or enduring cooperative relationships, fundamental for the survival of the social species. It is consequently intrinsically interpersonal and intersubjective.

(iii) frequently comprehends both an instinctive and a rational component.

(iv) trust judgements are influenced by and involve the coordinated activity of genes, hormones and brain functioning in the interaction with an external environment.

(v) trust decision making is always backed by a risk assessment except for the situations it is taken for granted.

4 The Concept of [Trust]: Defining Semantic Boundaries and Scope of Reference

[Trust] gets a rich semantically nuanced character according to the entities involved and according to the different situational contexts. Among many other possible examples, we can refer, for instance, to the trust in institutions—Townley and Garfield [39], i.e., believing that the people that give body to political, social.... institutions will always act according to the community's best interests; or trust in science, Oreskes [32], i.e., trusting the knowledge and skills of those that are responsible for scientific development or for the definition of scientific procedures, namely those that are health related. The introduction of new medical procedures or of new drugs is frequently viewed with suspicion by the potential end-users and sometimes not

[9] We lack evidence if this behaviour can also be found in organisms as plants.

even the certification by competent health authorities is capable of overcoming a generalised feeling of distrust, as it happens with large segments of world population relatively to vaccination in general and the Covid-19 vaccines in particular, cf. on this issue, the Vax-Trust Project.[10]

But the situational contexts that foster the emergence of these types of judgements are not limited to interpersonal, intersubjective relationships but can involve a relationship with other types of entities, namely artificial systems. For instance, Coeckelbergh [8] and Sullins [37] write about trusting robots and much has been written, in the recent years, about trustworthiness in what relates artificial intelligent systems, cf., The IEEE Global Initiative on Ethics of Autonomous and Intelligent Systems, IEEE, Ethically Aligned Design,[11]Trustworthy AI, cf. Chatila et al. [7], European Commission's key requirements for trustworthy AI,[12] Ethics Guidelines for Trustworthy AI [14] Tools for Trustworthy AI, OECD [30], to name just a few.

Language substantiates and objectifies a community's view of the world, reflecting its values, beliefs and feelings. We think one of the best ways to fully comprehend the nuanced character of the concept of [Trust], and consequently the nuanced character of the relationship established between [the human being who trusts] and [the entity trusted], is to identify the different semantic nature of the entities that may fit respectively into the first and second positions of the double slotted[13] argument structure defined by the concept:

According to the specific semantic nature of X and Y, we will be able to distinguish the two fundamentally different senses the concept assumes.

We will attempt a definition of [Trust^1] assumed as the basic sense, followed by a definition of [Trust^2].

(1) X trusts Y
x: human y: human

Definition X believes (based on the features that substantiate Y's positive image in X's mind) that, in a given situation, Y will act according to that belief converging and committing to X's interests/goals.

This general definition allows us to raise some possible working hypotheses:

(i) trust judgements involve risk assessment on the part of X (the one who trusts), except when trust is taken for granted (e.g., as in the case of individuals in a very close affective relationship.

[10] https://vax-trust.eu/. VAX-TRUST examines vaccine hesitancy as a broad societal phenomenon aiming to identify the societal factors that shape beliefs and attitudes towards vaccinations in current societies.

[11] https://standards.ieee.org/industry-connections/ec/ead1e-infographic/.

[12] https://www.aepd.es/sites/default/files/2019-12/ai-ethics-guidelines.pdf.

[13] For the sake of conciseness we will not analyse structures as: "X trusts Y with Z" or "X trusts Z to Y", which we consider as being a specification of the basic (1) X trusts Y.

(ii) there is not necessarily any apriori inherent vulnerability on the part of X, though the option for trusting may, in some cases, involve a variable degree of risk.
(iii) a trust relationship tends to be symmetrical as it develops.
(iv) there is no moral value necessarily attached to a trust judgement as exemplified in (1), e.g., no validation of Y's character in terms of honesty-I can be a criminal and still be totally trustworthy to my peers.
(v) though there is no moral value necessarily involved, transparency seems essential, to be paramount, to the building of X's coherent mental representation of Y, and a fundamental element a trust relationship feeds on.
(vi) it is when the behaviour of Y fails to meet X's image of Y and their assumptions that trust is irremediably broken. The negative impact on X caused by this failing is proportional to the intensity of their belief.

Let us look now to what happens to the meaning conveyed by [Trust] when the second slot of the argument structure is filled by a non-animate entity, namely a product. The term [product] corresponds here to any entity produced and commercialised, be it a good or a service.

(2) X trusts Y

x: human

y: an object (product)

Definition X expects Y to exhibit inherent quality and risk free character.

As we can observe the meaning conveyed by [Trust^1] and [Trust^2] are completely distinct. In the next section we will analyse the conditions under which [Trust^2] can be verified.

5 Trusting Products

5.1 General Framework

Contrary to [Trust^1], [Trust^2] results exclusively from a rational judgement on the part of X (the one who trusts) that relies on their own personal experience or, lacking that direct experience and/or competence to ground their judgement, will rely heavily on the judgement of other, namely on a certification issued by organisms, publicly acknowledged as qualified to do it, that declare the quality and consequently risk-free character of a product.

Stanford Golberg[14] points out that, in developed societies, consumers hold certain expectations relative to the quality of the products they buy/consume in a market

[14] https://www.kellogg.northwestern.edu/trust-project/videos/goldberg-ep-3.aspx.

regulated by rules that companies have to comply with, both in the production and in the marketing phases. Goldberg (ibid) highlights that these are expectations—normative expectations—and not predictions.

In fact, deciding to buy this or that product typically does not involve any kind of risk assessment on the part of the consumer, i.e. the anticipation of the potential risk involved by their option, because it is assumed (though sometimes wrongly) that all the products in the market are risk free.

The threshold defined by the society's normative expectations is socially and culturally determined. As Goldberg (ibid) refers, "if you happen to live in a community where being trustworthy is extremely highly valued and being untrustworthy is extremely disvalued, that will give individuals with whom you interact a great motive to be trustworthy, whereas if you live in other communities where those sorts of things aren't valued or perhaps not enforced with the same regularity, that can also affect other people's trustworthiness, and so have an impact on your perception of their trustworthiness".

In this sense, in a social context where consumers are educated to have a critical perspective on the nature of the products they use and on the personal and collective impact of their own consuming habits on individual well-being and on the environment, thresholds will be much higher and consequently what Goldberg refers as "normative expectations" will be much more detailed and demanding. When an educated consumer buys a product, for instance a dish washing machine, they take into account not just its cost, but its quality, efficiency and since the last decades the impact of its use on the environment.

The inherent quality of the product is sometimes immediately granted by the producer's/ brand's name[15] that due to the acknowledged quality of its products is capable of incorporating the features quality, efficiency and reliability into the brand's identity.

In developed countries, a particular set of products–machines–are accompanied by a certificate, issued by a competent entity, attesting that it complies to predefined rules—standards that guarantee their quality and reliability. In the European Community the Machinery Directive[16] "aims at the free market circulation on machinery and at the protection of workers and consumers using such machinery. It defines essential health and safety requirements of general application, supplemented by a number of more specific requirements for certain categories of machinery".

The Directive covers all stages along the processual pipeline[17] from the machine's conception in order to efficiently perform the required function in a safe and risk free way to its placing as a reliable and safe product in the market and the monitoring of its evolution in the interaction with consumers.

[15] The fact that for many consumers white label products, though cheaper, are viewed with suspicion is a proof of that.

[16] https://osha.europa.eu/en/legislation/directives/directive-2006-42-ec-of-the-european-parliament-and-of-the-council.

[17] https://osha.europa.eu/en/legislation/directives/directive-2006-42-ec-of-the-european-parliament-and-of-the-council.

According to the Machinery Directive (ibid.):

- Machinery must be designed and constructed so that it is fitted for its function, and can be operated, adjusted and maintained without putting persons at risk when these operations are carried out under the conditions foreseen but also taking into account any reasonably foreseeable misuse thereof.
- The manufacturer or his authorised representative should also ensure that a risk assessment is carried out for the machinery which he wishes to place on the market. For this purpose, he should determine which are the essential health and safety requirements applicable to his machinery and in respect of which he must take measures.
- The machinery must then be designed and constructed taking into account the results of the risk assessment.
- The manufacturer or his authorised representative should prepare a technical construction file which must be available on request and which demonstrates conformity of the machinery with the essential health and safety requirements.
- The manufacturers should retain full responsibility for certifying the conformity of their machinery to the provisions of this Directive. The CE marking should be fully recognised as being the only marking which guarantees that machinery conforms to the requirements of this Directive.
- All Member States shall take all appropriate measures to ensure that partly completed machinery can be placed on the market only if it satisfies the relevant provisions of this Directive.
- Member States should ensure their capacity to carry out effective market surveillance, taking account of guidelines developed by the Commission, in order to achieve the proper and uniform application of this Directive.
- Member States shall institute or appoint the competent authorities to monitor the conformity of machinery and partly completed machinery and define tasks, organisation and powers.
- Member States should provide for penalties applicable to infringements of the provisions of this Directive. Those penalties should be effective, proportionate and dissuasive.

According to the Directive two types of entities are assumed as responsible for the definition of the product's milestones: one is the producer/manufacturer, the other are the regulatory bodies at national and international level.

Throughout the Directive's lines emerges as fundamental (i) the extensive risk assessment every new system/machine has to undergo in the testbeds, as a prototype, and later, in real scenarios before it is in fact massively deployed, (ii) the certification of its efficiency and risk free character, (iii) its technological transparency stating precisely what are its components and how they articulate with each other in order to achieve its ultimate functional goal, (iv) the manufacturer's accountability, (v) the legal responsibility of Member States monitoring that all the products coming to the market are properly certified, enforcing this certification and applying penalities to all the eventual infringements.

Functional efficiency, extensive risk assessment, certification of the risk free character of the product in compliance with legal national and international regulatory frameworks are the basis for achieving consumer's trust.

5.2 *The Case of Artificial Intelligent Systems*

Computer development, ICTs, the internet, robotics/automation and AI have introduced new forms of interaction, of communication, of business making, of working… of being in the world. And as it has been happening since the introduction of electricity and the massive production of goods, consumers have been embracing technology, adopting and adapting, incorporating it acritically in their daily life, in a fascinated childlike- way, unaware of the eventual risks they are also exposing themselves to, because they just assume the objects, the services at their disposal to be risk free.

And though consumers generally do not question the safety or reliability of the technology they are using, blindly trusting its efficiency unless it somehow frustrates the expected performance failing to meet the user's goals, the concept of trust has been identified as a fundamental pillar for AIS acceptance, being at the centre stage of all discussions concerning the reliable and ethical[18] deployment of artificial intelligent systems, cf. Ethics Guidelines for Trustworthy AI [13].[19] In this report, the High-Level Expert Group on AI identifies three essential components which should be met in order to guarantee trustworthiness all along the systems' life cycle: (i) it should be lawful, complying with all applicable laws and regulations (ii) it should be ethical, ensuring adherence to ethical principles and values and (iii) it should be robust, both from a technical and social perspective.

As we have been pointing out, consumers tend to assume the products in the market as risk free and a substantial effort has to be made in order to develop in the end-users, from a very young age, a critical perspective on the products they consume daily in the different fields of their existence. Consumers of digital products, in particular, are generally completely alienated and, in most cases, use technology acritically. One of the very sensible areas relates how individuals in the course of their daily digital interactions, with multiple applications and interfaces, involuntarily feed a galaxy of commercial, economic, political…. interests with the data that results from their digital footprints. And in a world where information has extensively become data that can be appropriated and used by others, protecting the integrity of one's identity and the right to privacy become key issues. As Studley and Little [36] refer we live in

[18] Ethics, in this sense, (i) reflects on the nature of the changes technological innovation introduces in human reality by modifying behavioural patterns and lifestyles; (ii) identifies the eventual disruption caused on common values, accepted norms and relations and consequently the eventual harm to individuals, to the communities or to the environment; (iii) it defines a human-centred approach where human well-being and dignity as well as the respect for other species and the planet are prioritised.

[19] https://www.aepd.es/sites/default/files/2019-12/ai-ethics-guidelines.pdf.

a Surveillance Society where the ability to electronically monitor our movements has grown from being the purview of the State, to that of corporations such as Google and Facebook and perhaps soon, any company with the rights and money to access the data gathered by the multitude of sensors inside and outside our homes. As these authors point out (2021:77) "Given the fundamental changes which these technologies bring to our lives, it seems a moral necessity that citizens should be empowered to make informed choices about which uses of technology they welcome",

Education is consequently an urge and a priority to endow individuals and society as a whole not only with the know—how to handle digital tools skillfully but similarly to have a critical perspective on the nature and reliability of the artificial systems one interacts with, be them the kitchen robot or the intelligent vacuum cleaner, the robot at grandma's, the algorithm responsible for the pop ups of suggested readings in the smartphone or the war news transmitted by multiple channels. Summing up, education is necessary and urgent to raise awareness to the eventual risk factors present in the environment we are embedded in and that is no longer an exclusively physical social environment, but also assumes a virtual and/or a hybrid dimension where agency is performed by both human and artificial players.

It is absolutely essential and prior that consumers and all the other stakeholders involved in the processual pipeline that leads to the deployment of AIS (from the very initial concept designer to the ultimate end-users) are educated and/or trained to be capable of identifying the positive impacts a system can bring about to a particular domain, to an individual organisation and to specific context of use and also capable of assessing the eventual risks involved and the possible negative impacts on the individual, the community and on the environment.

Risk-aware developers will be better equipped to create more accurate and reliable technology overcoming the eventual limitations or hazardous malfunctioning of the still rudimentary intelligent technologies.

Addressing the topic of safety and well-being within the domain of AIS, the European Commission in "Building Trust in Human-Centric Artificial Intelligence" states that AI systems should integrate safety and security-by-design mechanisms to ensure that they are verifiably safe at every step, taking at heart the physical and mental safety of all concerned.

This entails seeking to maximise the benefits of AI systems while at the same time preventing and minimising their risks.

With the goal of assisting in this complex task, The OECD [30] produced what was called "a framework of tools for trustworthy AI". These tools, organised according to the categories, Technical, Procedural and Educational OECD ([30]:9–12), are described as instruments and structured methods that can be leveraged by others to facilitate their implementation of the AI Principles of Fairness, Transparency, Explainability and Robustness. According to the OECD proposal (ibd.), educational tools for trustworthy AI encompass mechanisms to build awareness, inform, prepare or upskill stakeholders involved in or affected by the implementation of an AI system. They include change management processes; capacity and awareness building tools; guidance for inclusive AI system design; and training programmes and educational materials. Depending on the implementation context, educational tools are designed to serve different audiences. They can be wide-reaching and open to the public at large or focus on a specific group affected by the implementation of an AI system, such as SMEs or workers.

As it should happen with all technology, AI systems have to be human-centric, i.e., they have to be conceived and designed as a tool that, respecting human specificities, can increase human performance exponentially, producing welfare and well-being. A tool that rests on a commitment to serve both the individual and the common good. In this sense, the theoretical grounding and the methodologies traditionally developed by the Ergonomics and Human Factors framework seem to be essential for designing artificial systems capable of safely co-existing and/or co-acting with human beings[20]in order to achieve sustainable development goals. Human Factors and Ergonomics, usually (EHF) a real human-centred approach, refers to the application of psychological and physiological principles concerning human beings to the engineering and design of products, processes, and systems. Taking into account the specificity of human physical architecture as well as the way human beings interact with their physical and social environments, EHF aims to enhance safety and comfort when this environment is populated with artificial systems, reducing risk factors and error and increasing productivity. We think the adoption of a Human Factors framework and methodology is fundamental in the design of both embodied and non-embodied artificial intelligent systems, in their testing and ulterior monitoring. This framework should also comprehend, as inherent human factors, the cultural and ethical dimensions always present in communities and that embody the sets of relevant values responsible for the regulation of the social interaction of the individuals belonging to a specific community.

We would like to conclude this section by quoting the German Standardisation Roadmap on Artificial Intelligence (2020):

> in a context of rapid technological change, it is essential that trust remains the bedrock of societies, communities, economies and sustainable development.

In our opinion the "trustworthiness" of AIS is granted by the three milestones present in the processual pipeline that characterises their development, these are Benchmarking, Standardisation and Certification.

5.3 The Role of Benchmarking, Standardisation and Certification

5.3.1 Benchmarking

Maybe because they try to replicate some features of human intelligence, artificial intelligent systems have sometimes been wrongly questioned for their Trustworthiness as if they were human beings. This figment of collective imagination, probably

[20] Cf on this purpose Fletcher [23], [24].

a fiction-generated fantasy, is particularly evident in the case of anthropomorphic robots functioning in a work setting or performing in the domestic space. We have tried to highlight in the previous sections the essential difference between [Trust]^1 and [Trust]^2 pointing out the distinctive features responsible for defining each of these senses. As referred, in what concerns machines, [Trust]^2 stands on a set of very objective conditions concerning the quality, efficiency and safety (no risk factor involved) of the product. This means that until it is launched into the market it has to be subjected to extensive testing according to previously accorded parameters and well-defined metrics in order to be later on certified that it complies with a set of standards and can consequently be acknowledged as a functionally reliable and safe product.

Benchmarking starts this process of evaluation that will ultimately lead to the introduction of a new product or product category in the market and in society. It involves assessing the performance/quality of a product (be it an object or a service) according to predefined functional parameters. It aims at:

- Qualitative Development
- Ongoing Improvement
- Upgrading or updating (according to the results).

According to Dai and Berleant [10] the following features are key in AI benchmarking:

1. Relevance: Benchmarks should measure important features.
2. Representativeness: Benchmark performance metrics should be broadly accepted by industry and academia.
3. Equity: All systems should be fairly compared.
4. Repeatability: Benchmark results should be verifiable.
5. Cost-effectiveness: Benchmark tests should be economical.
6. Scalability: Benchmark tests should measure from single server to multiple servers.
7. Transparency: Benchmark metrics should be readily understandable.

One of the major challenges in developing intelligent systems is clearly identifying the relevant aspects to be benchmarked and defining the clear metrics to do it. Artificial Intelligent systems have a broad scope of application. This broad scope comprehends different domains and multiple contexts of use that define their own "technological idiosyncrasies". The functionalities required are determined by the nature of the context (domain, setting, stakeholders involved and goals) and have to take into account the culture/values of that specific domain, organisation they will be embedded in. Either when testing a robot's interaction or the functioning of a single algorithm, tuning up performance according to the nature of the variables involved is fundamental. We illustrate how context plays an important role by very briefly

Fig. 1 ERL- Smart Cities Robotics Challenge; Episode 3-deliver coffee shop orders. *Credits to* ERL, Smart Cities Challenge (August 2018)[24]

describing benchmarking procedures in the context of RoboCup@Home.[21] We will follow in this description the authors[22] own words.

In the context of the European Robotics League—Consumer Service Robots (ERL Consumer), multiple benchmarks have been defined to evaluate the performance of people perception and object perception by service robots operating inside domestic environments.

The European Robotics League, funded by the European Commission to advance research, development and innovation in robotics and artificial intelligence, has run in three categories: consumer (previously service), emergency and professional (previously industrial) service robots. Presently teams from all three categories meet every two years in the ERL Smart Cities Robotics Challenge. Challenges take place according to "episodes" that embody a specific context of use. An episode defines a set of Task Benchmarks (TBM) each comprehending a variable number of Functional Benchmarks (FBM). Although the goal is to target specifically one functionality (or exceptionally, two), an effort to integrate the main functionality and the other ones might be required in order to complete the episode (Fig. 1).[23]

We would like to finish this section by referring to the need of incorporating ethics as one of the parameters to be benchmarked in AIS performance. As it happens with the other benchmarks, the context will determine which behaviours should

[21] https://www.eu-robotics.net/robotics_league/smart-cities/about/index.html.

[22] *Technical Committee: Meysam Basiri, Instituto Superior Técnico, Portugal, Pedro Lima, Instituto Superior Técnico, Portugal, Daniele Nardi, Sapienza University of Rome, Italy, Gianluca Bardaro, The Open University, UK, External Expert: Alessandro Saffiotti, Örebro University, Sweden.*

[24] http://sciroc.org/e03-deliver-coffee-shop-orders/.

[23] https://docs.google.com/document/d/122AS7SgOQe__Aj0V3fhoXb-766LgDYz9gZfcrG Y7jMM/edit

be validated as correct. We exemplify this, just with two examples, taken randomly, from a possible set that emerges in the context of a robot performing as a companion at grandma's. One is more systemic, relating to all the privacy issues relative to the data (images, text…) the robot incorporates in its daily interaction with a human being at their most private sphere—home. The other example relates to the behaviour(s) the robot should or should not have exhibited, given a particular social/cultural context. In this sense, let us imagine that one of TBMs is [Remind Grandma to take her pills] subsuming FBMs as [Fetching the pills], [Handing over the pills to Grandma] etc.

What should be the robot's behaviour when realising grandma's intentionally not taking any of these pills?

- Keep reminding her of the need to take the pill?;
- Reporting the fact to her doctor?;
- Reporting the fact to her relatives?.....
- Just "dropping" the case.

Still in this "competition format" we would like to refer to an identical approach and methodology followed in the Horizon 2020, Metrics project.[25]

METRICS aims at establishing a metrology-grade methodological framework for the evaluation of robotic systems in four priority areas.

The consortium defines an evaluation plan template for the METRICS competitions, including a methodology for the evaluation protocol, the evaluation tasks, the metrics used, the test facilities, test databases, participation rules and operating procedures.

The project covers four distinct domains: Healthcare, Inspection and Maintenance, Agri-Food and Agile Production. According to the specificity of each domain particular (TBMs) are defined, comprehending the set of (FBMs) from whose correct performance the realisation of the Task Benchmark depends.

For example in the Heart-Met Competitions (those dedicated to the Healthcare sector) a TBM is—[Fetch a glass of water]. In order to realise this TBM, the robot will have to go through a set of FBM, such as: FBM1: To understand fetching orders; FBM2: Navigate avoiding obstacles; FBM3: Object Detection…….

Also in the Metrics competitions we observe that the benchmarks focus on the correct functional realisation of a set of tasks that are determined by the specific domain and context of use, but that no eventual ethical benchmarks are identified and/or incorporated. As it happens with FBMs, these ethical benchmarks would be, in most cases, subsumed by each TBM.

This kind of awareness of the ethical and societal/cultural values that may emerge in distinct contexts of use and distinct domains is fundamental so that autonomous systems perform correctly in what concerns what are defined as human or environment fundamental values.

Under the headline "AI lacks Benchmarks and Consensus", the 2021 AI Index Report[26] refers that despite the growing conversation around AI ethics and related

[25] https://metricsproject.eu/.

[26] https://aiindex.stanford.edu/ai-index-report-2021/.

domains, the field significantly lacks benchmarks to measure or assess relationships between technologies and their impact on society. "Though a number of groups are producing a range of qualitative or normative outputs in the AI ethics domain, the field generally lacks benchmarks that can be used to measure or assess the relationship between broader societal discussions about technology development and the development of the technology itself".

5.3.2 Standardisation

Addressing the role of standards in our society, Alan Winfield mentioned[27] that standards are present in our lives on a 24 h—a day-basis. From those that regulate such trivial things as the quality of our toothbrush and toothpaste to all those that define the requirements for the products, services or processes present in most circumstances of our contemporary lived reality.

Laying the foundation for technical procurement and product development, they ensure the risk free character of the products we use or interact with and frequently also the interoperability among the equipment produced by different sources, introducing this way efficiency, sustainability and well-being in all walks of life.

According to Franklin [24] a standard is generally a document containing requirements representing consensus best practices in a given field. Industry standards are developed through voluntary cooperation by industry experts, as well as academic and government experts. The work is intended to support the success of industry through the establishing of predictable and best-practice requirements for that industry's product or service. It represents consensus best practices in a given field, developed according to a defined framework. Such a framework typically requires transparency, fairness, and balance among stakeholders, a balance between the interests and goals of producers and the needs of users.

Still according to Franklin (ibid.) there are two major approaches to standards development:

- The first aims to lead the development of technology, in order to shape it and create a market for it.
- The second states that experts cannot establish a consensus of best practices until a technology has been observed in use for some time. Machinery safety standards belong to this latter type.

In what relates to autonomous systems, as Franklin (ibd.,) points out, the risk of prematurely adopting a technology before it proves reliable is too great to permit rushing the development of consensus.

As we can observe in Fig. 2, the definition of a standard defines a loop. However, contrary to what it may look, this loop is not static but a very dynamic entity. This

[27] Keynote, at the International Conference on Robot Ethics and Standards 2019, London Southbank University.

Fig. 2 Production of a standard—stages of the process. *Credits to* BSI

means, it gets its definition in the context of an iterative process, that substantiates the process of innovation, development and commercialisation of a product.

At international level, ISO/IEC JTC 1/SC 42 "Artificial Intelligence" is the central body for AI standardization and is therefore responsible for the development and publication of international standards on AI systems. There are also a number of professional associations that publish corresponding specifications or recommendations. A considerable amount of work on AIS and AIS standardization has been developed by the IEEE subject-dedicated working groups. The IEEE 7001–2021 Standard for the Transparency of Automated Systems is just an example of such work.[28]

At the start of this document it is highlighted that:

- Compliance with IEEE standards does not constitute compliance to regulatory frameworks
- The use of IEEE std is wholly voluntary
- Users of IEEE std should rely on their own independent judgement.

The standard identifies all stakeholders directly or indirectly affected by the issue of Transparency and how they distribute along the different stages of the product lifespan. On the other hand, the System Transparency Specification (STS) defines the transparency requirements of an AI system for each of those stakeholders group

[28] https://media-exp1.licdn.com/dms/document/C4D1FAQE75SDL6s4O6Q/feedshare-document-pdf-analyzed/0/1649055234847?e=2147483647&v=beta&t=HUb8Wf1VEz8Ra1CI-oKiBKbW0OjLc6Fyxlq6MdegwIE.

while the System Transparency Assessment (STA) provides an assessment of the transparency of an AI system for each of those groups.

Standards and specifications promote the rapid transfer of technologies from research to application. According to the German Standardisation Roadmap on Artificial Intelligence, they form the basis for technical sovereignty and create a framework that promotes transparency and ensures security, quality and reliability which are, as we have been claiming, the substance of trustworthiness.

At a time a substantially consistent regulatory body is required for the effective and adequate production and deployment of embodied and non-embodied artificial intelligent systems, standardisation emerges as the tool capable of guaranteeing not only the intrinsic desired quality of those systems, but also the potential (desirable) universality that can be granted by standards across regional/economic borders.

Because of their fundamental role and because they are the result of the volunteer work of different stakeholders, standards should, in our opinion, be open access, i.e., contrary to the common practice, their consultation by all those interested should not demand any kind of payment/fee. This openness would guarantee (i) the aimed universality of best practices fostering the quality and risk-free character of products and consequently their reliability, (ii) the fair development of the global markets.

5.4 *Certification*

Quality criteria and test procedures are needed for the marketable conformity assessment and certification of AI systems.

Certification of a system's trustworthiness relies on high-quality test methods that take into account the particular nature of the variants that define its specific context of use, its setting, and the eventual impacts on the stakeholders involved. As we mentioned in the previous sections, it requires reliable and reproducible tests all along the procedural pipeline.

A process of certification that, as the German Standardisation Roadmap on Artificial Intelligence (GSRAI) refers, stands on the basis of concrete application cases. According to these, testing principles are to be tested, pilot tests carried out and standards derived which form the basis for AI certification and are to be introduced into international standardisation.

Still according to the GSRAI, the test methods to be developed serve on the one hand to confirm the assured properties of AI systems—product testing—and on the other hand to evaluate the measures taken by organisations providing AI systems—management system testing.

On the roles of standardisation and certification for product development and for fostering economy, The New Industrial Strategy for Europe ([18]:5) refers:

> the single market depends on robust, well-functioning systems for standardisation and certification. These help to increase the size of markets and provide legal certainty. Developing new standards and technical regulations, coupled with increased EU participation in international standardisation bodies, will be essential to boost industry's competitiveness.

The Union product safety legislation aims to ensure that products placed on the Union market meet high health, safety and environmental requirements and that such products can circulate freely throughout the Union. The sectorial legislation is complemented by the General Product Safety Directive (GPSD)[29] in line with the New Consumer Agenda of 2020.[30] The GPSD aims to: (i) update and modernise the general framework for the safety of non-food consumer products; (ii) preserve its role as a safety net for consumers; (iii) adapt the provisions to challenges posed by new technologies and online selling; and (iv) ensure a level playing field for businesses.

6 Conclusions

In the "White Paper on Artificial Intelligence: a European approach to excellence and trust", the EU Commission published its vision for a safe and responsible use of artificial intelligence. This document represents a first attempt to establish clear rules on what AI may and may not do, identifying and scaling the eventual risk factors involved, establishing red lines and suggesting approaches on how to enforce clarity and compliance with common accepted rules/standards.

The focus is on making AI systems capable of fostering the development of science, industry and society, while at the same time identifying and scaling the eventual potential risks associated with some of them, so that they can be avoided or minimised.

As we highlighted in the first section of this paper, evolving in an environment always demands, on the part of the cognitive entity, awareness of the risk factors eventually present in that environment and that may pose a risk or threat to their well-being or even their existence. We would like to add, however, that the tolerance to risk is also culture determined, i.e., different communities, at different times, may have different approaches to risk differentiating on how prone they are to subjecting their populations to risk factors for the sake of something else that is viewed as risk worthy. As the German Standardisation Roadmap to Artificial Intelligence refers, in practice, the fundamental evaluation of risk situations in society often leads to a quantification of the risk from an economic perspective. This economic assessment is, for instance, particularly evident in the management of Covid-19 crisis. Though it should be undisputed that human life is the highest good one could observe and still can see how the extent of the economic damage caused by maintaining quarantine is equated relatively to the risks to public health of losing restrictions.

We consider that having a human-centred approach to technological development implies that individual and collective well-being and safety will always be at the centre stage.

---

[29] https://ec.europa.eu/info/business-economy-euro/product-safety-and-requirements/product-safety/consumer-product-safety_en.

[30] https://eur-lex.europa.eu/legal-content/EN/TXT/?uri=CELEX:52020DC0696.

References

1. Aly A, Griffiths S, Stramandinoli F (2016) Metrics and benchmarks in human-robot interaction: recent advances in cognitive robotics. Cogn Syst Res. https://doi.org/10.1016/j.cogsys.2016.06.002
2. Amigoni F, Bastianelli E, Bonarini A, Fontana G, Hochgeschwender N, Iocchi L, Schiaffonati V (2016) Competitions for benchmarking. IEEE Robot Autom Mag 22(3):53–61
3. Arrow KJ (1974) The limits of organization. Norton, New York, NY, USA
4. Berg J, Dickhaut J, McCabe K (1995) Trust, reciprocity, and social-history. Games Econ Behav 10:122–142. https://doi.org/10.1006/game.1995.1027
5. Bjornskov C (2007) Determinants of generalized trust: a cross-county comparison. Pub Choice 130:1–21. https://doi.org/10.1007/s11127-006-9069-1
6. Bonsignorio F, Del Pobil AP (2015) Toward replicable and measurable robotics research. IEEE Robot Autom Mag 22(3):32–35
7. Chatila R, Dignum V, Fisher M, Giannoti F, Morik K, Russell S, Yening K (2021) Trustworthy AI. In: Braunschweig B, Ghallab M (eds) Reflections on artificial intelligence for humanity. Springer
8. Coeckelbergh M (2012) Can we trust robots? Ethics Inf Technol 14(1):53–60. https://doi.org/10.1007/s10676-011-9279-1
9. DIN, DKE (2020) German standardisation roadmap on artificial intelligence. https://www.din.de/resource/blob/772610/e96c34dd6b12900ea75b460538805349/normungsroadmap-en-data.pdf
10. Dai W, Berleant D (2019) Benchmarking contemporary deep learning hardware and frameworks: a survey of qualitative metrics. In: 2019 IEEE first international conference on cognitive machine intelligence (CogMI). IEEE, Los Angeles, CA, USA, pp 148–155. arXiv:1907.03626. doi:https://doi.org/10.1109/CogMI48466.2019.00029.
11. Damasio A (2005) Brain trust. Nature 435:571–572. https://doi.org/10.1038/435571a
12. Engelmann JM, Herrmann E (2016) Chimpanzees trust their friends. Curr Biol 26:252–256. Elsevier
13. Ethics Guidelines for Trustworthy AI. https://digital-strategy.ec.europa.eu/en/library/ethics-guidelines-trustworthy-ai
14. European Commission (2019) Ethics guidelines for trustworthy AI. High level working group on AI https://www.aepd.es/sites/default/files/2019-12/ai-ethics-guidelines.pdf
15. European Commission (2019) Building trust in human-centric artificial intelligence. https://ec.europa.eu/jrc/communities/en/community/digitranscope/document/building-trust-human-centric-artificial-intelligence
16. European Commission (2020) White paper on artificial intelligence: a European approach to excellence and trust. https://ec.europa.eu/info/publications/white-paper-artificial-intelligence-european-approach-excellence-and-trust_en
17. European Commission (2020) Report from the commission to the European parliament, the council and the European economic and social committee. Report on the safety and liability implications of Artificial Intelligence, the Internet of Things and robotics
18. European Commission (2020) A new industrial strategy for Europe, communication from the commission to the European parliament, the European council, the council, the European economic and social committee and the committee of the regions. https://eur-lex.europa.eu/legal-content/EN/TXT/PDF/?uri=CELEX:52020DC0102&from=EN
19. European Commission (2020) New consumer agenda. https://eur-lex.europa.eu/legal-content/EN/TXT/?uri=CELEX:52020DC0696
20. European Commission (2021) The 2021 coordinated plan on artificial intelligence is the next step in creating EU global leadership in trustworthy AI. https://digital-strategy.ec.europa.eu/en/library/coordinated-plan-artificial-intelligence-2021-review

21. European Commission (2021) General product safe directive. https://ec.europa.eu/info/business-economy-euro/product-safety-and-requirements/product-safety/consumer-product-safety_en
22. European Community (2019) Communication from the commission to the European parliament, the council, the European economic and social committee and the committee of the regions on building trust in human-centric artificial intelligence, Brussels, COM, 168 final
23. Fletcher S, Charalambous G (2021) Trust in human robot collaboration. In: Ferreira MIA, Fletcher SR (eds) The 21st century industrial robot: when tools become collaborators. ISCA series. Springer
24. Franklin C (2021) The role of standards in human-robot integration safety. In: Ferreira MIA, Fletcher S (eds) The 21st century industrial robot: when tools become collaborators. ISCA series. Springer
25. Heidegger M (1977) The question concerning technology and other essays. Garland Publishing Incorporated. New York, London. https://monoskop.org/images/4/44/Heidegger_Martin_The_Question_Concerning_Technology_and_Other_Essays.pdf
26. IEEE, Ethically aligned design, from theory to practice. Chair: Raja Chatila. https://standards.ieee.org/industry-connections/ec/ead1e-infographic/
27. IEEE, The IEEE global initiative on ethics of autonomous and intelligent systems. https://standards.ieee.org/wp-content/uploads/import/documents/other/ead1e-introduction.pdf
28. Johnson W (2007) Genetic and environmental in—fluences on behavior: capturing all the interplay. Psychol Rev 114:423–440. https://doi.org/10.1037/0033-295X.114.2.423
29. Le Roux M, Peterson S, Mougous J (2015) Bacterial danger sensing. J Mol Biol 427(23):3744–3753. Published online 2015 Oct 3. https://doi.org/10.1016/j.jmb.2015.09.018
30. OECD (2021) Tools for trustworthy AI: a framework to compare implementation tools for trustworthy AI systems. OECD Digit Econ Pap, n°312. https://www.oecd.org/science/tools-for-trustworthy-ai-008232ec-en.htm
31. OECD (2022) Enabling effective AI policies. https://oecd-events.org/2022-ai-wips/session/c6e2c2b1-bd7a-ec11-94f6-a04a5e7d3e1c
32. Oreskes N (2019) Why trust science? Princeton University Press, Princeton NJ
33. Pinker S (2005) Mind & language, vol 20, no 1 February 2005, pp 1–24. #Blackwell Publishing Ltd. 2005, 9600 Garsington Road, Oxford, OX4 2DQ, UK and 350 Main Street, Malden, MA 02148, USA. https://stevenpinker.com/files/pinker/files/so_how_does_the_mind_work.pdf
34. Riedl R, Javor A (2012) The biology of trust: integrating evidence from genetics, endocrinology, and functional brain imaging. J Neurosci Psychol Econ 2011 Am Psychol Assoc 5(2):63–91. https://doi.org/10.1037/a0026318
35. Righetti F, Finkenauer C (2011) If you are able to control yourself, I will trust you: the role of perceived self-control in interpersonal trust. J Pers Soc Psychol 100:874–886. https://doi.org/10.1037/a0021827
36. Studley M, Little H (2021) Robots in smart cities. In: Ferreira MIA (ed) How smart is your city?—technological innovation, ethics and inclusiveness. ISCA series. Springer
37. Sullins JP (2020) Trust in robots. Simon 2020:313–325
38. Todorov A, Baron SG, Oosterhof NN (2008) Evaluating face trustworthiness: A model based ap- proach. Soc Cogn Affect Neurosci 3:119–127. https://doi.org/10.1093/scan/nsn009
39. Townley C, Garfield JL (2013) Public trust. In: Makela P, Townley C (eds) Trust: analytic and applied perspectives. Rodopi Press, Amsterdam, pp 95–107
40. Wahlster W, Winterhalter C (2020) German standardisation roadmap on artificial intelligence, November 2021. DKE German Commission for Electrical, Electronic & Information Technologies of DIN and VDE

41. Welch M, Rivera R, Conway B, Yonskoski J, Lupton P, Giancola R (2005) Determinants and consequences of social trust. Sociol Inq. Wiley Online Library
42. Zak PJ, Kurzban R, Matzner WT (2004) The neurobiology of trust. Ann N Y Acad Sci 1032:224–227. https://doi.org/10.1196/annals.1314.025
43. Zeder M (2012) Pathways to animal domestication. Biodivers Agric Domest Evol Sustain 227–259

Informational Privacy and Trust in Autonomous Intelligent Systems

Roeland de Bruin

Abstract For a successful societal deployment of Autonomous Intelligent Systems (AIS), citizen's trust is crucial. The central theme of this Chapter is formed by the relationship between informational privacy protection, trust and acceptance of autonomous intelligent technology. This contribution is written from a legal perspective. Certain rules of the EU General Data Protection Regulation (GDPR) are analysed that may apply to AIS-solutions. It is investigated to what extent the application of these rules to AIS-solutions may influence citizen's trust that their informational privacy is well-protected.

1 Introduction

Rapid developments characterise Autonomous Intelligent Systems (AIS) technology. Gradually, systems are endowed with increasing autonomy,[1] allowing these to make decisions that could in the past only have been made by humans, based on increasing intelligence, i.e. means of perceiving and processing information—including natural

The themes and ideas comprised in this contribution are also encompassed in Roeland's dissertation, which has been published in 2022 [1].

R. de Bruin (✉)
Centre for Access to and Acceptance of Autonomous Intelligence, Utrecht University, Utrecht, Netherlands
e-mail: R.W.deBruin@uu.nl

Attorney-at-Law (Law & Technology), KienhuisHoving N.V., Enschede, Netherlands

[1] See *inter alia* Chopra, S. and White, L.F., *A Legal Theory for Autonomous Intelligent Agents*, Ann Arbor: University of Michigan Press 2011, p. 10 (autonomy) and Davies, C.R., 'An evolutionary step in intellectual property rights—Artificial intelligence and intellectual property', *Computer Law & Security Review* 27, 2011, p. 601–619 (intelligence); and Cock Buning, M. de, Belder, L., & Bruin, R.W. de, "Mapping the Legal Framework for the introduction into Society of Robots as Autonomous Intelligent Systems", in: Muller S., et al., *The Law of the Future and the Future of Law: Volume II*, TOAP 2012, p. 198, and further references there.

© The Author(s), under exclusive license to Springer Nature Switzerland AG 2022
M. I. A. Ferreira and M. O. Tokhi (eds.), *Towards Trustworthy Artificial Intelligent Systems*, Intelligent Systems, Control and Automation: Science and Engineering 102,
https://doi.org/10.1007/978-3-031-09823-9_3

language, and their ability to learn from "experience".[2] Self-driving cars; personalised news-feeds on social networks; targeted online advertising; robots in many forms and shapes with diverse purposes; and electronic judges all use Autonomous Intelligent Technology.[3]

Where innovators sometimes worry that certain aspects of law seem to stand in the way of technological developments in this field, a growing point of concern is that, once developed, broad societal adoption of AIS is uncertain.[4] Several models have been developed in literature that regard the adoption of innovation in general. These include for instance the Theory of Planned Behaviour,[5] and the Diffusion of Innovation theory.[6] One of the factors that might influence the willingness of consumers to adopt and endorse novel technology, is *trustworthiness* of an innovator,[7] and therewith corresponding *trust* by consumers.[8] Trust, in turn, is a multi-facetted concept, regarding the mental state of those who are to adopt technology, which Rousseau et al. define as "a psychological state comprising the intention to accept vulnerability based on positive expectations or behaviour of another".[9] Thus, when an innovator, or his innovation, is trustworthy in sense that a consumer perceives it to be "positive", such a consumer is likely to take certain risks for granted that are comprised in the innovation. Such "positiveness" may for instance follow from the risk-minimizing behaviour that is demonstrated by the innovator. Amongst many

---

[2] Ibidem.

[3] See for example Williams, A.P., and Scharre, P.D., *Autonomous Systems—Issues for Defence Policymakers*, The Hague: NATO Communications and Information Agency 2015, p. 4; Scharre, P., *Army of None: Autonomous Weapons and the Future of War*, New York: W. W. Norton & Company 2018; http://www.futureforall.org/transportation/future_of_transportation.htm; Strickland, E., "Autonomous Robot Surgeon Bests Humans in World First", *IEEE Spectrum* 4 May 2016, via http://spectrum.ieee.org/the-human-os/robotics/medical-robots/autonomous-robot-surgeon-bests-human-surgeons-in-world-first; Bhorat, Z., 'Do we still need judges in the age of Artificial Intelligence?', Opendemocracy.net, 9 August 2017, available via https://www.opendemocracy.net/transformation/ziyaad-bhorat/do-we-still-need-human-judges-in-age-of-artificial-intelligence.

[4] See for example on the relationship between innovation in the field of autonomous vehicles and societal acceptance thereof: Rezvani, Z, Jansson, J., and Bodin, J., "Advances in consumer electric vehicle adoption research: A review and research agenda", *Transportation Research Part D* 34 2015, pp. 122–136 (Rezvani, Jansson & Bodin 2015). [Refer to the other contributions in this book here].

[5] Ajzen, I., "The theory of planned behaviour". *Organizational Behavior and Human Decision Processes* 50 1991, pp. 179–211, cited in Rezvani, Jansson & Bodin 2018, p. 126–128.

[6] Rogers E.M. *Diffusion of Innovations*, New York: Free Press 2003 (5th edition), which builds upon the first edition that was published in 1995.

[7] Carter L., and Bélanger F., "The utilization of e-government services: citizen trust, innovation and acceptance factors", *Info Systems Journal* (2005), no. 15, pp. 5–25 (Carter and Bélanger 2005); Van Slyke et al. 2004.

[8] See also Rousseau, D.M., Burt, R.S., Sitkin, S., Camerer, C.F., "Not So Different After All: A Cross-discipline View of Trust", *The Academy of Management Review*, 1998, p. 395, cited Cock Buning, M. de & Senden, L. (eds.), *Private Regulation and Enforcement in the EU* p. 20, and their reference to Six, F., and Verhoest, K., (eds.), *Trust in Regulatory Regimes*, Cheltenham: Edward Elgar Publishing 2017, p. 3.

[9] Ibidem Rousseau et al., 1998, p. 395.

other things, privacy protection and data security are observed to play a role in the "perceived positiveness" by consumers of digital technology.[10] Regarding the adoption of autonomous vehicles for example, Dorothy Glancy points out a negative relationship between privacy and technology-adoption: "[w]ithout appropriate legal protections for privacy, autonomous vehicles could well meet "market resistance".[11] On a more positive note, she indicates that "assuring respect for user privacy is one of the best ways to foster trust and confidence in new technologies such as autonomous vehicles".[12]

Privacy protection may not only be one of the "ethical" norms to follow by innovators in the several AIS-fields, it is also strongly regulated. In the European Union for example, the General Data Protection Regulation (GDPR) entered into force in 2018, which entails strict rules for innovators, aimed at sturdy protection of the informational privacy of citizens.

In this contribution, I focus on the potential relationship between informational privacy, as regulated under the framework of the GDPR, and trust, which in turn may influence the adoption of AIS-technology.

Departing from the presumption that effective privacy protection positively impacts trust, and thus adoption of AIS-technology, the question is posed to what extent GDPR-norms are likely to contribute to citizen's trust regarding the outcomes of innovation in the field of Autonomous Intelligent Systems. Another presumption is that in virtually all forms of AIS-technology, personal data will be processed. As the envisaged size of this chapter would not allow for an all-encompassing review, I have chosen four GDPR-norms to investigate. After a general introduction of the GDPR and its aims in Sect. 2. I will focus on two of the ex-ante obligations regarding Technical and Organisational Measures that have to be implemented, and Data Protection Impact Assessments that need to be carried out before data processing activities may be commenced, in Sect. 3.1. Requirements regarding automated decision making are addressed in Sect. 3.2, and transatlantic transfer of personal data is addressed in Sect. 3.3. An answer to the research question is formulated in Sect. 4.

[10] See Carter & Bélanger 2005, p. 9, 21; Van Slyke, C., Belanger, F., and Comunale, C. L., "Factors influencing the adoption of web-based shopping: the impact of trust", ACM SIGMIS Database: the DATABASE for Advances in Information Systems Volume 35 Issue 2, Spring 2004, pp. 32–49 (Van Slyke 2004); Balboni, P., *Trustmarks: Third-party liability of trustmark organisations in Europe*, Tilburg (diss.) 2008, available online via https://pure.uvt.nl/ws/portalfiles/portal/1063399/Trustmarks.PDF p. 9–10; See also regarding autonomous vehicles: Fagnant, D.J., & Kockelman, K., "Preparing a nation for autonomous vehicles: opportunities, barriers and policy recommendations", *Transportation Research Part A*, 77 (2015), p. 177–178; and Glancy, D.J., "Privacy in Autonomous Vehicles", *Santa Clara Law Review* 2012, vol. 52, no. 4, p. 1225 (Glancy 2012).

[11] Glancy 2012, p. 1225.

[12] Ibidem, p. 1225–1226.

2 GDPR and Its Aims

The General Data Protection Regulation (GDPR) entered into force on May 25th 2018. From that date, the regulation directly applied in all EU jurisdictions. The GDPR builds upon earlier regulatory instruments, including the (now withdrawn) Data Protection Directive, as well as OECD Guidelines (which are still in force). Its aim is to protect the fundamental right of informational privacy of EU citizens, which is on a more general level enshrined in Article 8 of the Charter of Fundamental Rights of the European Union and Article 16(1) of the Treaty on Functioning of the European Union. The GDPR aims at eliminating obstacles to "flows of personal data within the Union",[13] whereas technological developments have to be facilitated, whilst ensuring at the same time a high level of privacy protection of EU citizens.[14] According to the 7th recital, this is amongst other things necessary for "creating the trust that will allow the digital economy to develop across the internal market", whereas "[n]atural persons should have control of their own personal data. Legal and practical certainty for natural persons, economic operators and public authorities should be enhanced".[15] Thus, the GDPR acknowledges that trust is necessary for economic activity, i.e. the deployment of novel (digital) technology, which can be fuelled by strong privacy protection and both legal and practical certainty regarding the norms that see to privacy protection.

3 GDPR & AIS—Case Examples

The GDPR applies to personal data processing activities which are carried out within the EU,[16] or which regard personal data of EU citizens.[17] *Personal data* are "personal data are defined as "any information relating to an identified or identifiable natural person" (*data subjects*),[18] even when data subjects are *indirectly* identifiable (for instance through identification numbers, location data or means of cross-linking or singling out which could lead to a data subject). Processing includes almost every activity that one could think of, and is defined as "any operation or set of operations which is performed on personal data or on sets of personal data, whether or not by automated means, such as collection, recording, organisation, structuring, storage, adaptation or alteration, retrieval, consultation, use, disclosure by transmission, dissemination or otherwise making available, alignment or combination, restriction, erasure or destruction".[19]

[13] Recital 10 to the GDPR.

[14] Recitals 6, 7, 10, 12, 13 to the GDPR.

[15] Ibidem recital 7.

[16] See Article 3(1) GDPR.

[17] See Article 3(2) GDPR.

[18] Article 4(1) GDPR.

[19] Article 4(2) GDPR.

Those who determine "purposes and means" of data processing activities, qualify as *controllers* under Article 4(7) GDPR. When a controller choses to outsource certain aspects of the data processing activity to a third party (who thus processes such data on behalf of the controller), that third party is defined as *processor* in Article 4(8).

These notions can be illustrated with the following two examples:

1. Autonomous vehicles will be equipped with self-learning systems that allow communication between those vehicles and other infrastructural objects, in order calculate how to drive safely, and to prevent collisions as much as possible.[20] The "live data" (including for instance continuous speed, geolocation data, information regarding steering and brake appliance et cetera and camera images) can be logged and stored for some time, for example to be used in the unfortunate event that the AV was implicated in an accident, or for software maintenance and user support purposes. Some of the recorded data, such as geolocation data and camera images, can identify the passengers of the AV (*data subjects*) and are *personal data* under the GDPR. The AV-producer company who decides which data to store, qualifies as *controller*. When he deploys a third party to maintain the software, and to physically store the processed data, that software company qualifies as *processor*.
2. AI technology had been used to study the proteins that form the SARS-CoV-2-virus, and likely elements thereof that would likely (not) mutate over time for the development of vaccines.[21] AI technology could now also be used to determine who will be vaccinated first,[22] and which vaccine would likely be most effective for specific patients,[23] on the basis of self-learning algorithms. The institution sending invitations to people who have been selected, on the basis of to the AI-based decisions regarding which parts of the population to invite (*data subjects*), and which vaccine to administer to them, qualifies as *controller*.

[20] See Kulk S., & Van Deursen, S. (eds.), *Juridische aspecten van algoritmen die besluiten nemen – een verkennend onderzoek*, UU/WODC 2020, p. 74–75; Surden H., & Williams, M.A., "Technological Opacity, Predictability, and Self-Driving Cars", *Cardozo Law Review* 2016 vol 36, no. 1, p. 121–181.

[21] See for example Malone, B., Simovski, B., Moliné, C. et al., "Artificial intelligence predicts the immunogenic landscape of SARS-CoV-2 leading to universal blueprints for vaccine designs", *Nature,* scientific reports 23 December 2020, 10, no. 22375; Walts, E., "What AI Can-and Can't-Do in the Race for a Coronavirus Vaccine", *IEEE Spectrum*, 20 September 2020, via https://spectrum.ieee.org/artificial-intelligence/medical-ai/what-ai-can-and-cant-do-in-the-race-for-a-coronavirus-vaccine (last accessed 15 January 2021).

[22] See the post by the De Montfort University, "Researchers argue that Artificial Intelligence can help decide who gets a Covid-19 vaccine first", 12 January 2021, via https://www.dmu.ac.uk/about-dmu/news/2021/january/researchers-argue-that-artificial-intelligence-can-help-decide-who-gets-a-covid-19-vaccine-first.aspx (last accessed 15 January 2021).

[23] See for example Ahuja, A.S., "The impact of artificial intelligence in medicine on the future role of the physician", *PeerJ*, 2019; 7: e7702, https://doi.org/10.7717/peerj.7702, who more generally observes on the basis of a literature study that AI can be used in order to decide which medicine to prescribe.

There are seven basic principles that need to be taken into account for every data processing activity. The *lawfulness, fairness and transparency* principle indicates that personal data may only be processed on a "lawful basis", in a fair and transparent manner in relation to the data subject.[24] Personal data may, regarding the *purpose limitation* principle only be collected and processed for specified, explicit and legitimate purposes, and may not be processed further in ways that are incompatible with the original purposes.[25] The *data minimisation* principle prescribes that data processing activities must be based on as little personal data as necessary.[26] Personal data must at all times be accurate (*accuracy principle*),[27] and kept up to date, and may only be stored as long as necessary to achieve the purpose (*storage limitation principle*).[28] Furthermore, the *integrity and confidentiality principle* dictates that appropriate security measures must be implemented to ensure the integrity and confidentiality of the processed data.[29] To conclude, the controller is always responsible for compliance with the GDPR-norms, and must be able to demonstrate his compliance on the basis of the *accountability principle*.[30]

These principles are further elaborated within the body of the GDPR. Before zooming in a bit further on certain obligations for *controllers*, it must be noted that non-compliance with the norms of the GDPR can have serious consequences. Not only is a controller (and sometimes also a processor) liable in tort towards any data subject (or a third party) who has suffered damage as a result of a GDPR-infringement, the norms can also be enforced by data protection authorities, who are able to impose administrative fines up to € 20.000.000,-, or 4% of the worldwide annual turnover (whichever is higher) of the non-complying controller or processor.

3.1 Ex-Ante Obligations: TOMs and DPIAs

TOMs

Controllers are *inter alia* obliged to make sure that the personal data they are processing,[31] remain intact, available and confident at all times, and must be able to demonstrate compliance. Regarding the security measures to take, Article 32 stipulates that controllers are obliged, "taking into account the state of the art, the costs

[24] Article 5(1)(a) GDPR.

[25] Ibidem, sub b.

[26] Sub c.

[27] Sub d.

[28] Sub e.

[29] Sub f.

[30] Article 5(2) GDPR.

[31] Other obligations include for example that a comprehensive record of data processing activities is kept (Article 30 GDPR); and that privacy-by-design and privacy-by-default principles must be implemented in organisations and innovations (for example software solutions) before data processing activities are begun with (Article 25 GDPR).

of implementation and the nature, scope, context and purposes of processing as well as the risks of varying likelihood and severity for the rights and freedoms of natural purposes", to implement technical and organisational measures (TOMs) "to ensure a level of security appropriate" to the respective risks. The same article presents some examples of TOMs to be implemented, including pseudonymization and encryption of personal data, and on a more general level, "the ability to ensure the ongoing confidentiality, integrity, availability and resilience of processing systems and services"; "the ability to restore the availability and access to personal data in a timely manner in the event of a physical or technical incident"; and the implementation of a "process for regularly testing, assessing and evaluating the effectiveness of technical and organisational measures for ensuring the security of the processing".[32] Where no more specific guidance is provided, it can be derived from the text of the article, that stricter measures must be taken when the processed data are of a more sensitive nature, such as *special category data*.[33]

Although Article 32(3) indicates that adherence to an approved code of conduct or certification mechanism "may be used as an element to demonstrate compliance", it must be noted that such approved codes or certificates do not exist to date. However, data protection authorities, including the European Data Protection Board (EDPB) and its predecessor the Article 29 Working Party, have adopted policy documents, from which more guidance for specific processing activities can be derived. Yet—these guidelines are no "hard" regulation, and can be overturned by a court.[34]

> The EDPB published guidelines for personal data processing through connected vehicles in 2020.[35] Regarding TOMs, they indicated the following TOMs[36]:

- "encrypting the communication channels by means of a state-of-the-art algorithm;
- putting in place an encryption-key management system that is unique to each vehicle, not to each model;
- when stored remotely, encrypting data by means of state-of-the-art algorithms;
- regularly renewing encryption keys;
- authenticating data-receiving devices;
- ensuring data integrity (e.g., by hashing);
- make access to personal data subject to reliable user authentication techniques (password, electronic certificate, etc.);"[37]

---

[32] Article 32(1)(a–d).

[33] The GDPR distinguishes between "normal" personal data and "special category data", which are defined in Article 9(1) as data: "revealing racial or ethnic origin, political opinions, religious or philosophical beliefs, or trade union membership, and the processing of genetic data, biometric data for the purpose of uniquely identifying a natural person, data concerning health or data concerning a natural person's sex life or sexual orientation".

[34] Which for instance recently occurred in the Dutch *VoetbalTV*-case: Rechtbank (District Court) Midden-Nederland 23 November 2020, ECLI:NL:RBMNE:2020:5111.

[35] European Data Protection Board, *Guidelines 1/2020 on processing personal data in the context of connected vehicles and mobility related applications*, version 1.0, 28 January 2020 (EDPB 01/2020).

[36] Ibidem, p. 19–20.

[37] Ibidem.

For vehicle manufacturers, the EDPB has further recommendations: "

- partitioning the vehicle's vital functions from those always relying on telecommunication capacities (e.g., "infotainment");
- implementing technical measures that enable vehicle manufacturers to rapidly patch security vulnerabilities during the entire lifespan of the vehicle;
- for the vehicle's vital functions, give priority as much as possible to using secure frequencies that are specifically dedicated to transportation;
- setting up an alarm system in case of attack on the vehicle's systems, with the possibility to operate in downgraded mode;
- storing a log history of any access to the vehicle's information system, e.g. going back six months as a maximum period, in order to enable the origin of a potential attack to be understood and periodically carry out a review of the logged information to detect possible anomalies."[38]

Regarding data collections including instantaneous speed (or other data that could reveal criminal offences), the EDPB furthermore recommends implementing strong security measures, such as: "

- implementing pseudonymisation measures (e.g. secret-key hashing of data like the surname/first name of the data subject and the serial number);
- storing data relating to instantaneous speed and to geolocation in separate databases (e.g. using state-of-the-art encryption mechanism with distinct keys and approval mechanisms);
- and/or deleting geolocation data as soon as the reference event or sequence is qualified (e.g. the type of road, day/night), and the storage of directly identifying data in a separate database that can only be accessed by a small number of people."[39]

Albeit there also are EDPB-guidelines for data processing "concerning health for the purpose of scientific research in the context of the COVID-19 outbreak",[40] these are not specific regarding which TOMs to implement, and the guidelines do not extend to data processing regarding decisions on whom to vaccinate first, with which vaccines – as illustrated in the second example in Sect. 1.3.

Thus, where some guidance for specific sectors regarding which TOMs to implement, others are left with the general principles set out in the GDPR, which implicates (even) less certainty for controllers and processors who operate in those sectors.

DPIAs

Another ex-ante compliance obligation to carry out a Data Protection Impact Analysis (DPIA). This obligation exists for controllers who envisage personal data processing

[38] Ibidem, p. 20.

[39] Ibidem, p. 29.

[40] European Data Protection Board, *Guidelines 03/2020 on the processing of data concerning health for the purpose of scientific research in the context of the COVID-19 outbreak*, 21 April 2002 (EDPB 03/2020).

activities which likely result in "a high risk to the rights and freedoms of natural persons".[41] Such high-risk processing activities include for example those in which "a systematic and extensive evaluation of personal aspects relating to natural persons which is based on automated processing, including profiling, and on which decisions are based that produce legal effects concerning the natural person or similarly significantly affect the natural person"; those which require large scale processing of *special category data* (including for instance health data and geolocation data); and those involving "systematic monitoring of a publicly accessible area on a large scale".[42]

Both examples sketched in Sect. 1.3, require a DPIA to be carried out. In the first example, *special category* geolocation data are constantly logged, and to the extent the AV-camera's also make recordings outside the AV, this also qualifies as "systematic monitoring". Automated decision making (regarding which people to vaccinate first, and which vaccine to administer) also necessitates a DPIA to be conducted.

Such DPIA should consist of at least four elements,[43] including in general the following: (1) A description of the processing operation and the purposes; (2) an assessment of the necessity and proportionality of the processing, related to the purposes; (3) an assessment of the risks for fundamental (privacy) rights of the data subjects as a result of the envisaged processing activity; and (4) an overview of the TOMs and other measures to minimise the assessed risks. Should the outcome of the DPIA be that there would be a "high" risk for the privacy of the data subjects concerned despite measures being taken to minimise such risks, the competent data protection authority must be consulted prior to the processing.[44]

Although there are many standards available for DPIAs, none of them is accredited or endorsed by the authorities. WP29 observed that many of those do correspond with the minimum requirements of the GDPR,[45] but it remains for the controllers to determine which standard to use, and to verify compliance with the legal norms. Uncertainties remain regarding the question whether or not an (envisaged) processing activity brings about "high risks" for the privacy of data subjects in specific cases, and thus whether to consult the authorities beforehand—or not.

3.2 *Automated Decision Making*

Data subjects have the right not to be subjected automated decision making (ADM). ADM is defined in Article 22(1) GDPR to comprise any "decision based solely on

[41] Article 35(1) GDPR.

[42] Article 35(3)(a–c).

[43] Article 35(7) GDPR.

[44] Article 36 GDPR.

[45] Article 29 Data Protection Working Party, "Guidelines on Data Protection Impact Assessment (DPIA) and determining whether processing is "likely to result in a high risk" for the purposes of Regulation 2016/679", 17/EN WP 248 rev.01.

automated processing, including profiling, which produces legal effects concerning him or her [a data subject], or significantly affects him or her".

> When no human would be involved in the decision that person A receives an invite for being vaccinated against SARS-CoV-2 than person B, and when it is automatically decided which vaccine is to be administered on the basis of health data of the respective person, this likely qualifies as ADM in sense of Article 22(1), as person A will likely have better chances of not contracting Covid-19 than person B (which significantly impacts both persons). Furthermore, should it for instance be that travel privileges are granted to vaccinated persons only, this also has legal effects for both persons, as for instance A will, and B won't be able to enter into a contract with a travel agency.[46]

ADM is prohibited *unless* (a) it is necessary for entering into, or the performance of a contract between the data subject and the controller who is using ADM; (b) ADM is explicitly authorised by specific EU, or Member State law; or (c) unless ADM is based on the data subject's *explicit consent*.[47] It must be noted that *explicit consent*, or a "substantial public interest on the basis of Union or Member State law", is always necessary to lift the ADM-prohibition when *special category data* are used in the decision making process.[48]

> As it will be very hard, if not impossible to obtain the *explicit informed consent* of the data subjects illustrated in the case above, and ADM is not necessary for the respective institutions to enter into a contract with data subjects, it seems that the envisaged ADM may not be used unless this is explicitly sanctioned by specific EU or Member State legislation. To my knowledge, no such legislation exists on the EU level at the moment.

Thus, it seems unlikely that the envisaged ADM process can easily be brought in compliance with the provisions of the GDPR.

3.3 Transatlantic Transfers of Personal Data

Personal data may not be "exported" to third parties outside the European Economic Area, unless Articles 44 and following of the GDPR are complied with, "in order to ensure that the level of protection of natural persons guaranteed by this Regulation is not undermined".[49] The GDPR stipulates three "transfer tools" that could be used to legalise data-export, being adequacy decisions,[50] binding corporate rules,[51] and appropriate safeguards, such as EC accredited standard contractual clauses.[52] Until

[46] Article 29 Data Protection Working Party, "Guidelines on Automated individual decision-making and Profiling for the purposes of Regulation 2016/679", 3 October 2017, revised on 6 February 2018, 17/EN WP 251 rev.01, p. 19–21.

[47] Article 22(2) GDPR.

[48] Article 22(4) GDPR.

[49] Article 44 GDPR.

[50] Article 45(1) GDPR.

[51] Article 47 GDPR.

[52] Article 46(1) GDPR.

recently, data exchange between the EU and the US could legally be based on the EU-US Privacy Shield, an adequacy decision endorsed by the European Commission.[53] The EU-US Privacy Shield framework was installed after its predecessor (Safe Harbor) had been annulled by the *Max Schrems I*-decision of the Court of Justice of the European Union (CJEU).[54]

As a result of Edward Snowden's revelations that US authorities could *inter alia* have access to personal data of EU citizens that were being processed in the US for reasons of national security, whereas EU citizens had no means of effectively safeguarding and enforcing their rights against those authorities, the Court held that the privacy of EU citizens could not be appropriately be guaranteed in the US. After the invalidation of Safe Harbor, the EU-US Privacy Shield was enacted, which held stronger safeguards for the protection of the privacy of EU citizens.

However, the Court observed that despite the "new" and stronger safeguards for EU citizens, US authorities could still enforce access to EU personal data on the basis of surveillance programs, while data subjects are not entitled to "actionable rights before the courts against US authorities".[55] Still, EU citizens were not observed to have sufficient cause of action against such agencies. All in all, the Court decided to invalidate the EU-US Privacy Shield too.

The consequences of the *Max Schrems II*-decision extend beyond the annulment of this specific adequacy decision. Also regarding the other mechanisms (appropriate safeguards such as the EC model clauses and binding corporate rules), it is observed that (*inter* alia) extra measures must be taken for countries in which the privacy rights of EU citizens are not appropriately safeguarded, in order to ensure that data subjects are provided "a level of protection essentially equivalent to that guaranteed within the EU, read in the light of the Charter of Fundamental Rights of the European Union".[56] The EDPB has interpreted the Court decision, and hinted which could, in several scenarios, constitute such extra measures.[57] Those supplementary measures include *inter alia* for the scenario in which an EU-based *controller* uses "hosting", "backup", or "transit" services from a US-based service provider (or other actor based in a "third country")—who does not need to access the stored personal data, that strong encryption measures are taken, using state-of-the-art techniques, without sharing the decryption keys with the *processor*.[58] In cases where a US-based (or other third country-based) recipient of personal data *does* need to access personal data on an aggregated or pseudonymised level (rather than access to personal data "in the

---

[53] Commission Implementing Decision (EU) 2016/1250 of 12 July 2016, pursuant to Directive 95/46/EC of the European Parliament and of the Council on the adequacy of the protection provided by the EU-U.S. Privacy Shield, *OJ* L 207/1.

[54] CJEU 6 October 2015, C-362/14, ECLI:EU:C:2015:650 (*Schrems I*).

[55] CJEU 16 July 2020, C-311/18, ECLI:EU:C:2020:559 (*Max Schrems II*), no. 192.

[56] CJEU *Max Schrems II*, para. 134; ruling no. 2.

[57] European Data Protection Board, *Recommendations 01/2020 on measures that supplement transfer tools to ensure compliance with the EU level of protection of personal data*, 10 November 2020 (EDPB R 01/2020); and *Recommendations 02/2020 on the European Essential Guarantees for surveillance measures*, 10 November 2020 (EDPB R 02/2020).

[58] EDPB R 01/2020, p. 22–23 (on backup and hosting); 24–25 (on transit).

clear"), data may only be shared that have been aggregated or pseudonymised on such a high level, that the data subjects cannot be specified, singled out or cross-referenced, without sharing the de-pseudonymisation keys with the importer (which will hardly ever be possible, as the EDPB notes).[59] Whereas the EDPB furthermore addresses data sharing with "protected recipients" and "split or multi party processing", it also lists situations in which *no* effective supplementary measures can be taken. These include scenarios in which a recipients in a third country would need to have access to personal data "in the clear", or situations in which data are shared between business entities based within the EU, and other entities in third countries.[60]

The *Max Schrems II*-decision entails that the data sharing between the EU-based AV-producer, and the US-based software developer, must be considered illegitimate – as the software developer needs to have access to the stored personal data for maintenance and user support purposes, and no effective supplementary measures can be taken in order to ensure the proper protection of the privacy rights of the data subjects.

4 Conclusions

The case examples above allow several general observations to be made which can be used to construe answers to the question to what extent GDPR-norms are likely to contribute to citizen's trust regarding the outcomes of innovation in the field of Autonomous Intelligent Systems. First, it seems an uneasy task for *controllers* who are to deploy AIS-technology, to establish which technical and organisational measures must be implemented, as the regulated norms are open and vague. Some guidance is provided by the EDPB for certain appliances (in connected vehicles for example), but all-encompassing guidance is not available. Furthermore, the guidance provided by the data protection authorities remains "just policy", and can in practice be overturned in case law. Whereas regulatory specification is envisaged by the legislator for instance through accredited codes of conduct or certification mechanisms, such instruments are not yet in place. A similar observation can be made regarding compliance with the norms regarding data protection impact assessments: it remains uncertain when a "high" risk necessitates for example prior consultation with a data protection authority. Thus, this leaves *controllers* in an uncertain position regarding their compliance with the GDPR-norms, whilst non-compliance can be enforced both through civil liability actions and high administrative fines.

The second observation regards automated decision making. Where ADM-processes use special category data, and where *explicit informed consent* cannot be obtained and where specific Union- or Member State legislation is non-existent, automated decision making is prohibited under the GDPR.

The third observation regards data processing activities outside the European Economic Area. Where personal data are sent to a country which does not offer appropriate safeguarding of the privacy rights of EU-citizens, "supplementary measures"

[59] Ibidem, p. 23–24.

[60] Ibidem, p. 26–27.

must be taken. However, when a recipient which is based in the US, needs access to the personal data "in the clear" there are no supplementary measures that can be deemed effective. In other words: continuing such data sharing activities would be incompliant with the GDPR. This implicates that many data processing activities that have long been commonplace, have now been declared illegitimate. A solution is not provided yet, and needs to be found by the legislator. Until that time, continuing incompliant data sharing activities, lead to civil liability- and public enforcement risks.

The reviewed GDPR-provisions are *inter alia* aimed at the creation of citizen's trust that their personal data are well protected by parties who process these data. However, compliance with the norms by *controllers* is not so easy, as it sometimes seems hard to establish when sufficient action is taken (in the case of TOMs and DPIAs), or when existing data processing models appear to have become invalid, and no easy solutions can be implemented to re-validate these. All in all, GDPR compliance by AIS-providers thus seems to involve significant time and effort, whilst ensuring that compliance is reached cannot be established with certainty. How innovators in their role of *controllers* will respond to legal uncertainties and red-tape that follow from the reviewed norms, cannot be established without further research. However, one can imagine two "extremes" as to how innovators would, regarding the case examples, would respond thereto. Some innovators may put in all available efforts to reach maximum compliance—perhaps motivated by the fundamental rights to be respected, perhaps (also) motivated by the threats of enforcement. Others, who may not be scared-off by those risks, or for other, more practical reasons, could choose not to comply with the norms. Several "shades of grey" are also imaginable—which may even be the most likely scenarios: reasonable efforts are made, but as it is not certain when overall compliance is reached, within certain limits.

Assuming that maximum compliance can hardly be reached, and thus maximum protection of citizen's informational privacy rights, it is likely that there is room for improvement regarding the factors comprised in the GDPR that may influence citizen's trust. To be more specific, further clarification of the open norms regarding *inter alia* TOMs and DPIAs would significantly contribute to increasing legal certainty. Whereas the GDPR explicitly provides possibilities to regulate such specifications through codes of conduct,[61] or certification mechanisms,[62] further co-regulatory activity (in which the respective stakeholders would participate, as envisaged in the GDPR) can be recommended.

Furthermore, it is recommendable that the current de facto risks that could result from the *de iure* prohibition of transatlantic personal data processing, are properly addressed by the legislator. It can be imagined that currently implemented data processing practices which involve data-export to third countries such as the US cannot be remigrated swiftly—if at all—back to the EU, and that transatlantic processing activities are not seized immediately (or even in the longer run). This would however implicate (a) non-observance of the privacy rights of EU citizens,

[61] Article 40 GDPR.

[62] Article 42 GDPR.

and (b) hefty enforcement risks for data-exporters. It is strongly recommendable that a political solution with the respective third countries is sought, as solving these problems would extend beyond what can reasonably be achieved by innovators, especially when no viable EU-based alternatives can be found for data processing activities through services of parties based in third countries.

Reference

1. Bruin RWD (2022) Regulating Innovation of Autonomous Vehicles: Improving Liability & Privacy in Europe (diss), Amsterdam: deLex

Ethical Risk Assessment for Social Robots: Case Studies in Smart Robot Toys

Alan F. T. Winfield, Anouk van Maris, Katie Winkle, Marina Jirotka, Pericle Salvini, Helena Webb, Arianna Schuler Scott, Jaimie Lee Freeman, Lars Kunze, Petr Slovak, and Nikki Theofanopoulou

Abstract Risk Assessment is a well known and powerful method for discovering and mitigating risks, and hence improving safety. Ethical Risk Assessment uses the same approach, but extends the scope of risk to cover ethical risks in addition to

---

A. F. T. Winfield · A. van Maris (✉)
Bristol Robotics Laboratory, University of the West of England, Bristol, UK
e-mail: anouk_van_maris@hotmail.com

A. F. T. Winfield
e-mail: alan.winfield@brl.ac.uk

K. Winkle
School of Electrical Engineering and Computer Science, KTH Royal Institute of Technology, Stockholm, Sweden
e-mail: winkle@kth.se

M. Jirotka · P. Salvini · H. Webb · A. S. Scott
Department of Computer Science, University of Oxford, Oxford, UK
e-mail: marina.jirotka@cs.ox.ac.uk

P. Salvini
e-mail: pericle.salvini@cs.ox.ac.uk

H. Webb
e-mail: helena.webb@cs.ox.ac.uk

A. S. Scott
e-mail: arianna.schuler.scott@cs.ox.ac.uk

J. L. Freeman
Oxford Internet Institute, University of Oxford, Oxford, UK
e-mail: jaimie.freeman@oii.ox.ac.uk

L. Kunze
Department of Engineering Science, Oxford Robotics Institute, University of Oxford, Oxford, UK
e-mail: lars@robots.ox.ac.uk

P. Slovak · N. Theofanopoulou
Department of Informatics, King's College London, London, UK
e-mail: petr.slovak@kcl.ac.uk

N. Theofanopoulou
e-mail: nikki.theofanopoulou@kcl.ac.uk

© The Author(s), under exclusive license to Springer Nature Switzerland AG 2022

M. I. A. Ferreira and M. O. Tokhi (eds.), *Towards Trustworthy Artificial Intelligent Systems*, Intelligent Systems, Control and Automation: Science and Engineering 102,
https://doi.org/10.1007/978-3-031-09823-9_4

safety risks. In this paper we outline Ethical Risk Assessment (ERA), and set ERA within the broader framework of Responsible Robotics. We then illustrate ERA, first with a hypothetical smart robot teddy bear (RoboTed), and later with an actual smart robot toy (Purrble). Through these two case studies this paper demonstrates the value of ERA and how consideration of ethical risks can prompt design changes, resulting in more ethical and sustainable robots.

1 Introduction

Risk assessment is a well-known method for discovering and mitigating risks, and hence improving safety. Ethical Risk Assessment is not new; it is essentially what research ethics committees do [1]. But the idea of extending the scope of safety risk assessment of intelligent systems to encompass ethical risks is new. Given the growing awareness of the ethical risks of intelligent systems in recent years, ethical risk assessment offers a powerful method for systematically identifying and mitigating the ethical, societal and environmental risks associated with the use of robots and artificial intelligence (AI).

In Sect. 2 we first define ethical risk assessment (ERA) with reference to British Standard BS8611. Then, in Sect. 3, we determine whether this standard can be used as a guideline for ERA by testing it on a hypothetical smart robot teddy bear we call RoboTed. A fictional robot is used here as it allows us to evaluate a broad range of technological features that may not all be available in existing robots. In order to determine whether ERA is applicable to real-world robots, Sect. 4 goes on to evaluate an existing smart robot toy called Purrble.[1] We believe this to be the first work applying ethical risk assessment to a commercial smart robot toy. The paper concludes with a comparison of the assessment of the two robot toys and provides an appraisal of both the benefits and limitations of ERA. This paper is an extended version of a paper presented at ICRES 2020 [2].

2 Ethical Risk Assessment

Risk Assessment is a process that typically has three stages:

1. identify and analyse potential events (hazards) that may cause harm to individuals, property, and/or the environment;
2. make judgments on the acceptability and likely impact of the harm arising from exposure to the hazard (risks), then

[1] https://purrble.com/.

3. determine what steps should be taken to mitigate those risks and hence minimise or eliminate possible harms.

Standards for risk assessment are well established in safety critical systems. ISO 14971:2007 Application of risk management to medical devices, for instance, provides requirements and guidance for risk assessment for medical devices. And ISO 12100:2010 *Safety of machinery—Risk assessment and risk reduction* sets out requirements for performing risk assessments, notably including risk analysis focused on hazard identification.

Almost certainly the world's first explicitly ethical standard in robotics is BS8611-2016 *Guide to the ethical design and application of robots and robotic systems.* "BS8611 is not a code of practice, but instead guidance on how designers can undertake an ethical risk assessment of their robot or system, and mitigate any ethical risks so identified. At its heart is a set of 20 distinct ethical hazards and risks, grouped under four categories: societal, application, commercial and financial, and environmental. Advice on measures to mitigate the impact of each risk is given, along with suggestions on how such measures might be verified or validated" [3]. Societal hazards include, for example, anthropomorphisation, loss of trust, deception, infringements of privacy and confidentiality, addiction, and loss of employment.

BS8611 defines an ethical harm as "anything likely to compromise psychological and/or societal and environmental well-being", an ethical hazard as "a potential source of ethical harm"; and an ethical risk as the "probability of ethical harm occurring from the frequency and severity of exposure to a hazard" [4]. Ethical risk assessment thus extends the scope of risk assessment to include ethical harms, hazards and risks (in addition to physical harms, hazards and risks).

Psychological safety is a well-known topic in Human-Robot Interaction (HRI) studies, although often overlooked [5]. It is concerned with the reduction of stress and anxiety caused by a robot's appearance and behaviour (such as size, shape, adherence to social norms) and the potential impact of robots that can emulate human or animal characters (such as feelings and expression of emotions).

Among the most relevant guidelines on psychological safety for robotics and AI currently available, or in draft, are: British Standard BS8611, a guide to the ethical design and application of robots [4], the European Parliament report on recommendations to the Commission on Civil Law Rules on Robotics [6] and the IEEE standards project P7014™ on ethical considerations in emulated empathy in autonomous and intelligent systems [7].

Ethical Risk Assessment is a fundamental part of the practice of Responsible Robots, which we define as "the application of Responsible Innovation [8] in the design, manufacture, operation, repair and end-of-life recycling of robots, that seeks the most benefit to individuals and society and the least harm to the environment" [9]. We would expect ERA to be undertaken within a framework of responsible innovation—such as EPSRC's AREA framework[2]—and alongside ethically aligned [10] and values-based design [11].

---

[2] https://epsrc.ukri.org/research/framework/area/.

Fig. 1 Teddy

In the following two sections, ERA will be applied to both fictional and existing smart robot toys, to determine whether ERA can be applied to a broad range of technological features in smart robot toys.

3 Case Study with a Fictional Smart Toy: RoboTed

Consider a hypothetical smart toy in the form of a teddy bear named RoboTed. RoboTed is a simplified version of the robot Teddy from the 2001 movie *A.I. Artificial Intelligence*, directed by Steven Spielberg,[3] to better reflect today's capabilities (see Fig. 1).

This fictional RoboTed is designed to:

1. recognise its owner, learning their face and name, turning its face toward the child,
2. respond to physical play such as hugs and tickles,
3. tell stories, while allowing a child to interrupt the story to ask questions or ask for sections to be repeated,
4. sing songs, while encouraging the child to sing along and learn the song, and
5. act as a child minder, allowing parents to remotely listen, watch and speak via RoboTed.

For these functionalities, RoboTed is based upon the following technology:

1. It is an Internet (WiFi) connected device,
2. It has cloud-based speech recognition and conversational AI (chatbot) and local speech synthesis,
3. Its eyes are functional cameras allowing the robot to recognise faces,
4. It has motorised arms and legs to provide it with limited baby-like movement and locomotion—not walking but shuffling and crawling, and

[3] As a tribute to Stanley Kubrick.

5. It has touch sensors which allow it to respond to physical play.

As a worked example we now consider the ethical hazards and risks of RoboTed, under the four categories of: physical (safety) risks, psychological risks, privacy and security risks, and environmental risks.

3.1 Physical Risks

Tripping—as RoboTed crawls on the floor, it has the potential to become a trip hazard. A mitigation strategy might be to have RoboTed make an audible crawling sound when it is moving, to alerts users (particularly adults) to its presence.

Battery overheating—There is a risk that defective batteries or battery chargers can overheat or *in extremis* catch fire. In mitigation RoboTed should be designed to make use of low-risk consumer rechargeable batteries rather than high-risk Li-Ion batteries. In addition parents should be advised to supervise battery re-charging.

3.2 Psychological Risks

Addiction—RoboTed might be so compelling that it leads to a child playing obsessively with RoboTed and neglecting his or her family [12]. This also increases the risk of emotional distress should RoboTed's behaviour change or fail in any way (e.g. if the facial recognition was to fail and no longer recognises the child). A mitigation strategy might be to explore the addition of a RoboTed 'needs to sleep' function, as a way of limiting length of play times.

Deception—There is a risk that the child comes to believe that RoboTed has feelings for her [13]. To mitigate this risk we could design the chatbot to avoid language that suggest feelings, so that RoboTed never says things like 'I like you' or 'Why are you sad?'

Over trusting by the child—Building on deception there is a risk that the child cannot tell whether RoboTed is operating autonomously or is in the child minder mode. This may result in her sharing sensitive information in the belief that no one else will hear it, when actually her parents are watching and listening. The reverse is also true, in that she may share something she wishes her parents to know but is too embarrassed to raise face to face, when actually the robot is operating autonomously and her parents are *not* listening. Mitigation strategies would be concerned with making the mode of operation as obvious to the child as possible, for example only using RoboTed's speech synthesis when in autonomous operation.

Over trusting by parents—The risk here is that parents become over reliant on RoboTed's child minder function [14]. The risk and its consequences are so great as to suggest the child minder function should be removed altogether.

The Uncanny Valley—The Uncanny Valley can lead to a fearful reaction when a robot is close to but not 100% lifelike [15]. The risk of this is probably low with RoboTed, both because RoboTed is not human-like at all, and children are already familiar with teddy bears. However, the risk should be explored by engaging children in early trials of RoboTed, and if the uncanny valley reaction is demonstrated it might be mitigated by, for instance, equipping the robot with a cartoon voice.

3.3 Privacy and Security Risks

Weak security—Weak security could lead to malicious hackers gaining access to RoboTed's sensors and control functions. This could be very frightening for a child and her parents. To reduce the risk we need to implement strong encryption of the communications between RoboTed and the cloud, alongside best practice password protection to make it very hard for hackers to guess the password.

Privacy—Here the risk is that personal data, including images and voice recordings of children (and the house they live in) are stolen. One way of reducing this risk would be to ensure that personal data sent to the cloud is deleted immediately after it has been used.

Lack of transparency—The risk is that if there were an accident in which RoboTed harmed a child, that could be either physical or psychological harm, it would be very difficult to investigate what happened to cause the accident unless the robot keeps a data log of its actions and responses. This is a serious risk and to mitigate the risk a secure data logger (*ethical black box*) needs to be built into RoboTed [16]. The data would be stored locally, and only the most recent few hours of data would need to be saved.

3.4 Environmental Risks

Unsustainability of materials—Here the risk is that the robot uses unsustainable or high carbon materials. To mitigate this risk we could use materials (e.g. RoboTed's fur) from sustainable sources. We could also avoid plastics by, for instance, using wood for RoboTed's skeleton.

Unrepairability—This leads to the risk that the robot's lifetime is limited because faults cannot be repaired or parts replaced. This risk can be minimised by designing RoboTed for ease of repair, using replaceable parts as much as possible (especially the battery). Additionally, RoboTed's manufacturers should provide a repair manual so that local workshops can fix most faults.

Unrecyclability—All products will eventually come to the end of their useful life, and if they cannot be repaired or recycled we risk them being dumped in landfill. To mitigate this risk, RoboTed should be designed to make it easy to recycle parts.

Ideally after these parts have been recovered for recycling, the remaining materials are biodegradable.

3.5 *Discussion of ERA for RoboTed*

The evaluation of potential risks of RoboTed, as summarized in Table 1, has demonstrated the value of ethical risk assessment. It has shown that a focus on ethical risks can:

- suggest new functions, such as 'RoboTed needs to sleep now',
- draw attention to how designs can be modified to mitigate some risks,
- highlight the need for user engagement, and
- reject some product functionality as too risky.

Testing ERA with a fictional robot has value as it allows us to cover a broad range of functionalities and hazards. However, for a more complete understanding of the application and limitations of ERA, we next consider an existing smart robot toy.

4 Case Study with Existing Smart Toy: Purrble

Our second case study focusses on an existing commercial product emerging from HCI research, named Purrble (see Fig. 2). Purrble is a small, interactive plush animal that has been specifically designed to provide an in-situ emotion regulation support to children, which is specifically 'child-led': this means that the design assumes that no training should be required for the child (or their adults) for the emotion regulation effects to occur. A full description of the design process and the intervention theory of change can be found in Theofanopoulou et al. [17] and Slovák et al. [18].

In summary, the robot is presented to children as a vulnerable creature, which is often anxious (indicated by a fast heartbeat-like vibration) that calms down when cuddled. The expectation is—and empirical data from previous studies suggests—that by 'soothing' the robot, the child down-regulates themselves; that children will start seeking the Purrble when they themselves are distressed, and that these repeated engagements can constructively shift emotion-regulation processes in the family.

Technologically, the robot is very simple, as presented in Fig. 3:

1. The only modes of communication are vibration patterns (fast to slow heartbeats, and a purr when soothed), as well as several squeaks added for the commercial units (growls and chirps, no language)
2. Sensors consist of
   (a) touch sensors on the back and sides, detecting (any) kind of touch, and
   (b) a gyro, tuned so that rapid movements 'scare' the creature as does turning it 'upside down'.

Table 1 Ethical risk assessment of RoboTed

Hazard	Risk	Level	Mitigation
Physical risks			
Tripping	User(s) trip over RoboTed when it is crawling on the floor	M	Audible crawling sound to alert users to its presence
Battery overheating	Defective batteries or battery chargers can overheat or *in extremis* catch fire	M	Design to make use of consumer rechargeable batteries rather than high-risk Li-Ion batteries
Psychological risks			
Addiction	Child plays with RoboTed obsessively and neglects family	M	Explore 'RoboTed needs to sleep now' function
Deception (of child)	Child believes that RoboTed has feelings (for her)	M	Design chatbot to avoid language that suggests feelings
Over trusting (by child)	Child cannot distinguish mode of operation	H	Notification when child minder mode activated, uses parents' voice rather than RoboTed's voice
Over trusting (by parents)	Parents come to rely on the childminder function	H	Remove the childminder function
The Uncanny Valley	Child becomes fearful of Robot	L	Use cartoon voice; engage children in early user trials
Privacy and security risks			
Weak security	Malicious hackers gain access to RoboTed's sensors and control function	H	Implement strong encryption together with best practice password protection
Privacy	Personal data, including images and voice recordings of child are stolen	M	Put in place auditable measures to ensure personal data is deleted immediately
Lack of transparency	Lack of data logs makes it hard or impossible to investigate accidents	H	Build a secure local data logger into RoboTed
Environmental risks			
Unsustainability (of materials)	Robot uses unsustainable or high carbon cost materials	M	Use materials (e.g. RoboTed's fur) from sustainable sources, avoiding plastics
Unrepairability	Robot's lifetime is limited because faults cannot be repaired or parts replaced	M	Design for ease of repair with replaceable parts—especially battery
Unrecyclability	End of life robots are dumped in land fill	M	Design for ease of recycling parts and materials

Note Risk level: (H)igh, (M)edium or (L)ow

Fig. 2 Purrble, taken from www.purrble.com (accessed 10–01–2020)

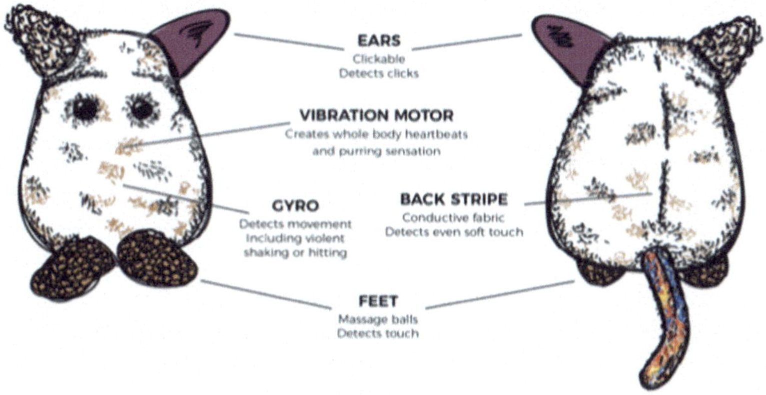

Fig. 3 Technological features of Purrble (taken from [17])

3. No network connection or data collection is present on the commercial units; the software is flashed onto the boards and thus immutable.
4. No movement functionalities are present, apart from the vibration.
5. The 'AI' of the robot is a very simple finite-state machine, representing a linear counter that changes vibration in response to previous state and outputs, by simple 'points' addition/subtraction (e.g., touching back $==+1$ point; toy upside down $=-20$ points). As a result, the empirically observed assumptions of 'liveness' or 'emotions' are brought in by emotion projection of those using the toy.

Before presenting our ERA for Purrble, it should be highlighted that some risks are equally applicable to many toys or artifacts (e.g. a choking hazard if eyes fall off). We contacted the company Sproutel, the developers of Purrble, to ask about their approach towards these hazards. They responded that Purrble meets international toy safety standards, and that they obtained certificates to confirm compliance with requirements on product safety. Therefore, such risks will not be addressed in this work and the ERA will solely focus on risks unique to *robot* toys.

4.1 Physical Risks

Injury—the battery pack of Purrble is hidden underneath a layer of fur. As this layer is relatively thin, Purrble's underside is relatively hard. This can result in physical harm if the Purrble is dropped on a person or thrown by the child (if she is upset perhaps). This risk can be mitigated by either providing a denser layer of fur over the battery pack, or relocating the battery pack less close to the surface of the toy.

Hygiene—due to its functionality, it is likely that Purrble will be shared, for example between siblings or in a classroom. Currently, the fur of Purrble is not washable, presenting a hygiene risk.[4] To reduce this risk the Purrble should be updated such that the fur can be taken off the robot and washed.

4.2 Psychological Risks

Influence relationships between people—the use of Purrble may impact relationships between people. For example, it may disrupt home dynamics (e.g. in contexts such as trauma, fostered or adopted families), or exacerbate already existing tensions. As there are many possible scenarios for this hazard with different risk levels, a thorough analysis on human relationships—and different circumstances in which Purrble can be used—is needed in order to develop mitigation approaches. We do not address this any further here as it goes beyond the scope of this study.

Dependency—if the child becomes too dependent on Purrble, there is a risk of increased anxiety and/or stress if Purrble is unavailable when needed. This could occur for example, if the batteries die, or another child is using Purrble, or it is lost or stolen. This can be mitigated by providing an indicator when the batteries are running low, and ensuring a child's time with Purrble is limited to prevent her becoming too dependent on it.

Overreliance—people might become too reliant on the supportive function of Purrble, using it as a 'quick fix' during situations for which it was not designed and not addressing deeper underlying issues. This risk is currently low as Purrble is still new to the market, but may increase with a larger user base. The risk can be mitigated by stressing the importance of carefully considering whether Purrble is the correct support system to use, case by case.

Nurturing Machines—children may feel compelled to look after Purrble [19]. The main function of Purrble is that it needs to be supported to calm down. The risk of

[4] Highlighted by the 2019–2021 Covid-19 pandemic.

feeling compelled to care has been mitigated by designing Purrble such that it is not too needy, resulting in relatively short interaction times.

Impact on interaction skills—intensive use of Purrble may lead to the child forgetting about, and interacting less with, other living creatures (e.g. classmates, siblings or pets). Replacement by Purrble may become the new norm (a phenomenon also known as environmental amnesia [20]), resulting in a decrease in development of interaction skills. This risk can be mitigated by limiting interaction time with Purrble.

Animal welfare—due to the autonomy and interactive abilities of Purrble it may be more likely to be considered a 'companion' for a child that otherwise may have been a pet. Currently, this risk is low as Purrble is still new. However, this risk highlights the advantage that Purrble can be used in situations in which pets are disallowed.

Negative symbolism—the goal of Purrble is to support emotion-regulation [17]. However, Purrble can become a symbol for mental health challenges which can lead to incorrect conclusions regarding the person's mental health, bullying or unwanted emotional conversations that people may prefer to keep private. The level of this risk will increase once people become more familiar with Purrble and can be mitigated by considering the way Purrble is introduced.

4.3 Privacy and Security Risks

Currently, Purrble does not incorporate any technologies that may raise privacy and/or security risks. It is not hackable as it is fully autonomous and does not require any connections; is it not open source nor does it store data. If any of these factors change potential privacy and security risks will need to be considered.

4.4 Environmental Risks

The environmental risks for Purrble are the same as for RoboTed (Sect. 3.4).

4.5 Discussion of ERA for Purrble

The evaluation of potential risks of Purrble, are summarized in Table 2. The ERA of Purrble provided several interesting insights, indicating that ERA is a useful tool for existing smart robot toys as well as the fictional RoboTed. Comparing the ERAs for Purrble and RoboTed, we see that the risks are less distributed over the four risk categories for Purrble than they are for RoboTed. This was expected, as RoboTed presents a broad range of functionalities that allowed for assessment of all four categories, where the functionality of Purrble is constrained to providing psychological support.

Table 2 Ethical risk assessment of Purrble

Hazard	Risk	Level	Mitigation
Physical risks			
Injury	The battery pack may injure a person if the robot is used forcefully	L	Increase the thickness of the layer of fur covering the battery pack
Hygiene	Spread virus/disease	M	Update device such that fur can be taken off and cleaned
Psychological risks			
Disturb relationships	Disrupt home dynamics, exacerbate existing tensions	L/M/H	See description
Dependency	Increased anxiety if Purrble is unavailable when needed	M	Decrease interaction time with Purrble, provide indicator for battery level
Overreliance	Purrble used as 'quick fix' instead of addressing underlying issues	L/M	Ensure Purrble is not the only option considered to address issues
Nurturing machines	The child feels responsible for Purrble	L	Ensure behaviour displayed by Purrble is not too 'needy'
Impact interaction skills	Interaction skills are impacted due to less interactions with other living beings	M	Ensure interaction time with Purrble is limited and interaction with humans or animals encouraged
Animal welfare	Purrble is substituted for a pet	L	Similar to the hazard 'overreliance' Purrble should not be considered as the only option to address issues
Negative symbolism	Purrble becomes a symbol of mental health challenges, resulting in unwanted interactions	M	Consider how Purrble is advertised
Privacy and security risks: none determined for the current version of Purrble as it uses no technologies that can result in risk			
Environmental risks			
Unsustainability (of materials)	Robot uses unsustainable or high carbon cost materials	M	Replace unsustainable materials with materials from sustainable sources
Unrepairability	Robot's lifetime is limited because fault cannot be repaired or parts replaced	M	Adapt design for ease of repair with replaceable parts—exchange battery for rechargeable one
Unrecyclability	End of life robots are dumped in land fill	M	Adapt design for ease of recycling parts and materials

Note Risk level: (H)igh, (M)edium or (L)ow

Purrble has been developed, and is advertised, as a research-based product. However, this ERA highlighted that there is a risk of the toy becoming a source of anxiety, flipping its original purpose. This reveals a difference between the ERAs for Purrble and RoboTed. Psychological risks for RoboTed were directly linked to the functionality of the smart robot toy, whereas some of the psychological risks for Purrble were indirect consequences of the use of a smart robot toy that provides psychological support (e.g. the concern regarding negative symbolism).

It should be noted that Purrble is still in the early stages of production. If sales take off, other aspects besides the ones presented in this ERA should be considered. For example, currently the electronics used are single use as they are cheaper to produce. If demand for the smart toy increases, the sustainability of the design should be considered, and whether the lifetime of the robots could be extended through providing options for repair.

Furthermore, even though Purrble has been developed to support children with emotion-regulation, research may indicate that the toy can be beneficial to other users. In this case, an additional ERA would be required to determine whether there are new risks that do not apply to children as a user group or have been overlooked. For example, Purrble seems to be opening up conversations that allow the parents and their child to talk about intimate personal emotional concerns. However, such conversations require different approaches depending on the user group. It is essential to carefully consider how Purrble is best introduced to specific user groups to ensure that both its functionality and the expectations of the user are understood.

5 Discussion and Conclusions

This paper has shown the value of ethical risk assessment of smart robot toys through both fictional and real-robot case studies. The assessments have indicated that attention to ethical risks can:

- suggest new functions,
- draw attention to potential design modifications to mitigate some risks,
- highlight the need for user engagement,
- reject product functionality as too risky, and/or
- indicate potential future issues, highlighting the need for periodic reassessments.

ERA is however, not guaranteed to expose all ethical risks. It is a subjective process which will only be successful if the risk assessment team are prepared to think both critically and creatively about the question: what could go wrong? As Vallor et al. [21] write, design teams must develop a "habit of exercising the skill of *moral imagination* to see how an ethical failure of the project might easily happen, and to understand the preventable causes so that they can be mitigated or avoided".

The ethical hazards and risks set out in BS8611 are an excellent starting point, but the standard does not provide an exhaustive taxonomy of ethical hazards, encompassing all domains of robotics. The RoboTed case study has identified several additional ethical hazards, some of which are specific to social robots, including the Uncanny Valley, weak security, lack of transparency (for instance the lack of data logs needed to investigate accidents), unrepairability and unrecyclability. Additionally the Purrble case study has indicated that there are psychological risks which do not follow directly from the use of the device, but indirectly, for example a disturbance of the relationship between siblings, which may lead to bullying. Such risks are not easily identified solely from the use of BS8611 and were exposed by co-authors with experience in many areas such as AI, responsible innovation, ethics, social robotics, human centred computing and psychology. Evaluating and quantifying psychological risks is especially difficult given that there are no agreed measures for hazards such as over trusting, the Uncanny Valley or when the child becomes too dependent on the smart robot toy. Assessment is further complicated by the likelihood that cultural and individual differences may lead to lower risks of psychological harm for some individuals than others. For these reasons design teams cannot rely on their own judgement and instead should engage with potential users from across the full range of age, gender and ethnic diversity, and seek guidance from psychologists and/or social scientists, both to ask the user group the right questions and interpret their responses.

Given also that ERA is not a one-time process but one that should iterate throughout a product life-cycle, good practice suggests that in-house assessments undertaken early in the design process would be shared with user groups during later iterations as the product undergoes user trials. During these iterations, the carbon footprint required for the production of the smart robot toy should be established.

Consider also what impact ERA might have on cost and user acceptance. If all mitigation strategies were applied, these might diminish the marketability of the product—which may not be a bad thing if the product is unethical (e.g. the doll 'My Friend Cayla' banned in Germany for privacy reasons [22]). However, whether companies will be able to make an adequate return on investment if they adhere to the suggested mitigations remains an open question. Note that there would be considerable value in a quality mark for responsibly designed and developed robot toys, similar to the FRR quality mark developed by the Foundation for Responsible Robotics that is currently in its pilot phase.[5]

In summary, ethical risk assessment is a powerful and essential addition to the responsible roboticist's toolkit. ERA can also be thought of as the opposite face of robot accident investigation [9], seeking—at design time—to prevent risks becoming accidents.

Acknowledgements The work of this paper has been conducted within EPSRC project RoboTIPS, grant reference EP/S005099/1 *RoboTIPS: Developing Responsible Robots for the Digital Economy*. The authors are also grateful for the comments of the anonymous reviewers of ICRES 2020. The

[5] https://responsiblerobotics.org/quality-mark/.

work of this extended paper was partially supported by the Assuring Autonomy International Programme, a partnership between Lloyd's Register Foundation and the University of York.

References

1. Bernabe R, van Thiel G, Raaijmakers J, van Delden J (2012) The risk-benefit task of research ethics committees: an evaluation of current approaches and the need to incorporate decision studies methods. BMC Med Ethics. https://doi.org/10.1186/1472-6939-13-6
2. Winfield AFT, Winkle K (2020) RoboTed: a case study in ethical risk assessment. Presented at the 5th international conference on robot ethics and standards (ICRES 2020), 28–29 September 2020. arXiv preprint: arXiv:2007.15864
3. Winfield A (2019) Ethical standards in robotics and AI. Nat Electron 2:46–48. https://doi.org/10.1038/s41928-019-0213-6
4. BSI (2016) BS8611:2016 Robots and robotic devices, Guide to the ethical design and application of robots and robotic systems. British Standards Institute
5. Lasota P, Fong T, Shah J (2017) A survey of methods for safe human-robot interaction. Found Trends Robot 5:261–349. https://doi.org/10.1561/2300000052
6. Delvaux M (2017) Report with recommendations to the commission on civil law rules on robotics. European Parliament, 27
7. IEEE P7014 Standard for Ethical Considerations in Emulated Empathy in Autonomous and Intelligent Systems. https://standards.ieee.org/project/7014.html. Accessed 7 Jan 2021
8. Stilgoe J, Owen R, Macnaghten P (2013) Developing a framework for responsible innovation. Res Policy 42(9):1568–1580. https://doi.org/10.1016/j.respol.2013.05.008
9. Winfield AF, Winkle K, Webb H, Lyngs U, Jirotka M, Macrae C (2020) Robot accident investigation: a case study in responsible robotics. arXiv preprint arXiv:2005.07474
10. IEEE (2019) The IEEE global initiative on ethics of autonomous and intelligent systems. Ethically aligned design: a vision for prioritizing human well-being with autonomous and intelligent systems, 1st edn. Tech. Rep. IEEE Standards Association
11. Spiekermann S, Winkler T (2020, May 12) Value-based engineering for ethics by design. Available at SSRN: https://ssrn.com/abstract=3598911 or http://dx.doi.org/10.2139/ssrn.3598911
12. Sharkey N, Sharkey A (2010) The crying shame of robot nannies. Interact Stud 11:161–190. https://doi.org/10.1075/is.11.2.01sha
13. Sparrow R, Sparrow L (2006) In the hands of machines? The future of aged care. Mind Mach 16:141–161. https://doi.org/10.1007/s11023-006-9030-6
14. Feil-Seifer D, Mataric M (2011) Socially assistive robotics. IEEE Robot Autom Mag 18:24–31. https://doi.org/10.1109/mra.2010.940150
15. Moore R (2012) A Bayesian explanation of the 'Uncanny Valley' effect and related psychological phenomena. Nat Sci Rep 2:864. https://doi.org/10.1038/srep00864
16. Winfield AFT, Jirotka M (2017) The case for an ethical black box. In: Gao Y, Fallah S, Jin Y, Lekakou C (eds) Towards autonomous robotic systems. TAROS 2017. Lecture notes in computer science, vol 10454. Springer, Cham. https://doi.org/10.1007/978-3-319-64107-2_21
17. Theofanopoulou N, Isbister K, Edbrooke-Childs J, Slovák P (2019) A smart toy intervention to promote emotion regulation in middle childhood: feasibility study. JMIR Mental Health 6:e14029. https://doi.org/10.2196/14029
18. Slovák P, Theofanopoulou N, Cecchet A et al (2018) I just let him cry …. Proc ACM Hum Comput Interact 2:1–34. https://doi.org/10.1145/3274429
19. Turkle S (2007) Authenticity in the age of digital companions. Interact Stud 8:501–517. https://doi.org/10.1075/is.8.3.11tur

20. Kahn PH Jr, Severson RL, Ruckert JH (2009) The human relation with nature and technological nature. Curr Dir Psychol Sci 18(1):37–42. https://doi.org/10.1111/j.1467-8721.2009.01602.x
21. Vallor S, Green B, Raicu I (2018) Ethics in technology practice. The Markkula Center for Applied Ethics, Santa Clara University
22. Madnick S, Johnson S, Huang K (2019) What countries and companies can do when trade and cybersecurity overlap. Harvard Business Review, p 4

Practical and Open Source Best Practices for Ethical Machine Learning

Jeroen Franse, Violeta Misheva, and Daniel S. Vale

Abstract The acknowledged operational, ethical, legal and governance risks involved in applying Machine Learning (ML) have generated a need for a clear and thoughtful repository of best practices on how to responsibly govern, manage and implement "responsible ML". The Foundation for Best Practices in Machine Learning (a non-profit foundation) seeks to promote responsible ML through creating an open-sourced, freely accessible repository of best practices and associated guides. Its model and organisational guides look at both the technical and institutional requirements needed to promote responsible ML. Both blueprints touch on subjects such as "Fairness & Non-Discrimination", "Representativeness & Specification", "Product Traceability", "Explainability" amongst other topics. Where the organisational guide relates to organisation-wide process and responsibilities (i.e. the necessity of setting proper product definitions and risk portfolios); the model guide details issues ranging from cost function specification and optimisation to selection function characterization, from disparate impact metrics to local explanations and counterfactuals. It also addresses issues concerning thorough product management. These guidelines have been developed principally *by* senior ML engineers, data scientists, data science managers, and legal professionals *for* ML engineers, data scientists, data science managers, compliance professionals, legal practitioners, and, more broadly, management. The Foundation's philosophy is that (a) context is key, (b) responsible ML starts with prudent MLOps and product management, and (c) responsible ML needs to be supported by all aspects of an organisation's structure.

J. Franse (✉) · V. Misheva · D. S. Vale
The Foundation for Best Practices in Machine Learning, Leiden, The Netherlands
e-mail: jeroen.franse@fbpml.org

V. Misheva
e-mail: violeta.misheva@fbpml.org

D. S. Vale
e-mail: daniel.vale@fbpml.org

© The Author(s), under exclusive license to Springer Nature Switzerland AG 2022

M. I. A. Ferreira and M. O. Tokhi (eds.), *Towards Trustworthy Artificial Intelligent Systems*, Intelligent Systems, Control and Automation: Science and Engineering 102,
https://doi.org/10.1007/978-3-031-09823-9_5

1 Introduction

In recent years, the fast and often reckless adoption of machine learning (ML) has led to applications with undesirable consequences, including (but not limited to): racial bias in facial recognition [1]; gender bias in recruiting [2] and credit risk [3]; and physical harm in medicine [4, 5]. Of these examples, only [4] and [5] were stopped before being deployed. Many other examples have received plenty of media and public attention, for an overview see [6, 7]. This has also spurred many initiatives directed at addressing these issues at an organizational and institutional level [8]. These guides rely principally on theoretical principles by which an ethical ML solution should abide. However, there is little to no guidance on how to apply these principles in practice or on what it means to comply with them [9].

This fact has been our main source of motivation for the work presented here: despite the existing efforts, it remains profoundly difficult for developers of ML applications to obtain practical guidance on developing ML responsibly; neither on what it means to be responsible or who carries what responsibility when it comes to ML development.

2 Ethical Machine Learning

The Foundation for Best Practices in Machine Learning [10] aims to fill in that gap. Our mission is to 'Champion ethical and responsible machine learning through open-source best practices and free public knowledge'. Our members are volunteers and have diverse backgrounds: data scientists, machine learning engineers, statisticians, legal professionals, academics, and communications experts. The Foundation is a non-profit organization which means that all our work is open-sourced under the Creative Commons Attribution Licence under which it may be freely reused by anyone, even for commercial purposes, as long as the Foundation is appropriately credited.

The way we propose to decrease the unwanted and unfair consequences of ML (a complete prevention is perhaps not realistic) relies on three main pillars:

- Contextualisation: every case is different and the solution needs to carefully consider the specific situation.
- Prudent MLOps (Machine Learning Operations) and Product Management: enable and conduct thorough management of the ML model lifecycle and the product lifecycle.
- Deep organisational support: the organisation cannot burden the development team with the sole responsibility of ethical product development. Instead, it should provide them with tools, policies, and resources.

Let's take as an example a government institution building an application for welfare fraud detection. This application will have very different fairness, explainability and redress and appeal requirements depending on how an individual flagged as 'high-fraud-risk' by the application is treated. Does it mean that (a) a welfare check for someone in a vulnerable financial situation arrives a couple of days too late, (b) someone from a marginalized group is placed on a register or under surveillance, or (c) someone receives help in the form of coaching.

If we want to ensure responsible products, one needs to follow prudent Product Management and ML Operations practices. These will ensure, for example, that the correct stakeholders (for example, affected citizens) are consulted throughout important decision points, and that the application is recalibrated when straying out of acceptable operating bounds. Lastly, the development team alone, or the government official using the application cannot bear the sole responsibility for the ethical development or usage of the application. Instead, the institution should provide all necessary policies, tools, resources and all required support. For instance, a clear chain of escalation will provide ways for the development team to obtain additional expertise and resources when needed.

Following these principles will result in fairer and safer ML applications, lowering the overall risk. At the same time, this also leads to better performing and robust models, and a higher rate of models and products being successfully deployed to production.

It is quite difficult to develop a framework which is universal and normative while remaining practical and generic. It is nearly impossible to come up with an exhaustive list of instructions about what one should do in any given situation. Even if that was possible, very often there is not a single correct answer when we deal with ethical dilemmas. Therefore, we face the challenge of developing a framework which is both practical and generic. We try to be as detailed as possible until we hit either normative issues or lose general applicability. We don't give a prescriptive answer; instead we help identify and detect the issues, and direct the users to methods and tools to use for a possible resolution.

3 Best Practices in Machine Learning

3.1 Definition and Scope

Our Best Practices (BP) are at the core of our Foundation. They are a pair of documents: one about organisational issues and one about technical issues. We aim to be complete and that makes the set of BP rather large. Neither does following them strictly result in a guaranteed problem-free and ethical by design ML product. In other words, they are not a silver bullet.

For these reasons we have also provided two ways to help one make their way around the BP and apply them:

1. Firstly, we have written user guides. These are short, easily accessible documents that help one navigate the BP and get started with them.
2. Second, we have developed an open-source portal, which everyone can use to contribute to the BP. It also has room for supporting material such as examples, elaborations, references and literature. We are starting to add this material, but we expect that the community can provide significant contributions here as well.

Finally, we expect that there is a lot of opportunity for an ecosystem of tools to develop around the BP; tools that may help to navigate, document or administer their usage, for example.

The BP are not limited to an industry or a specific team within an organisation. They are suitable for different audiences with varying levels of technical expertise (data scientists, engineers, developers but also legal and compliance professionals). They are also suitable for all types of organisations, regardless of the maturity, domain, size or potential social impact of the company. Of course, when we need to go into technical detail we do not shy away from that; but at the same time, the BPs can help build a common language and understanding across all stakeholders in an organisation.

3.2 Technical and Organisational BP

We created two BP:

1. The Technical BP for a single product;
2. The Organisational BP for an entire organisation.

Both documents are based on the same categorization of component subjects (see below). These subjects are mediated in the Technical BP through the phases of model lifecycle management, whereas they are mediated through policies in the Organisational BP. You can think of policies as any set of guidelines, tools and protocols that the organisation provides to their product teams to help apply the Best Practices in their own context.

Through these shared subjects and the policies, the two BP documents are tied together (Fig. 1).

We've made the following clustering of subjects in the BPs:

- Product Management
- Fairness and Non-Discrimination
- Data Quality
- Representativeness and Specification
- Performance Robustness
- Monitoring and Maintenance
- Explainability
- Security
- Safety

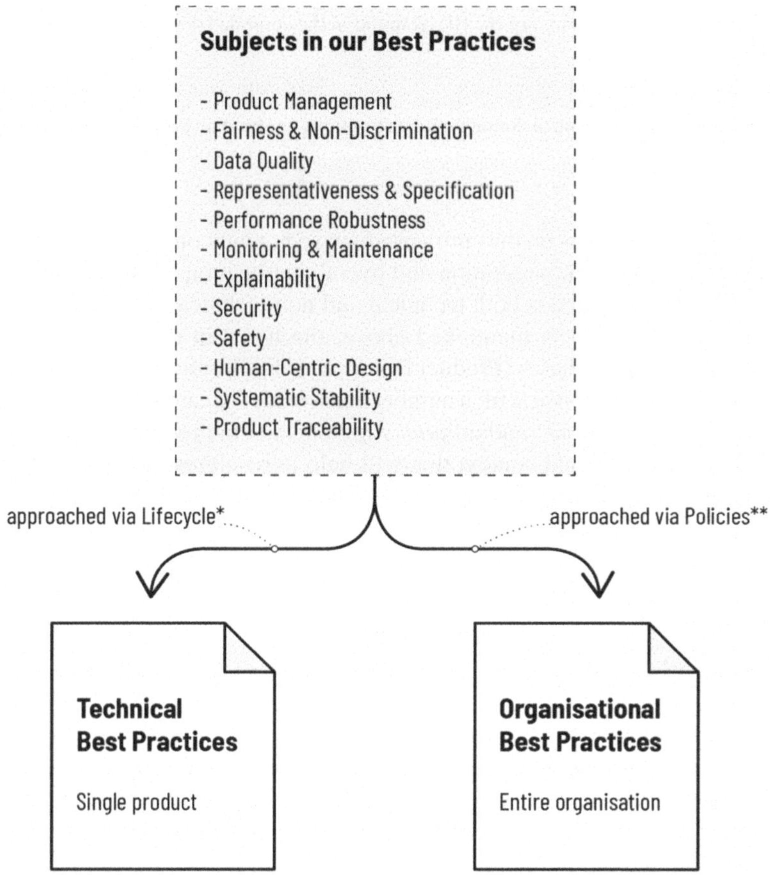

Fig. 1 The BPs and their subjects of focus

- Human-centric Design
- Systemic Stability
- Product Traceability.

This particular clustering of subjects is based on the type of perspective one needs to take while working on the associated issues and, therefore, aids to practical application.

For example, Fairness and Non-Discrimination is about preventing biases that are measurable in a more or less straightforward way; whereas Representativeness and Specification is about trying to look ahead at what might cause biases that will be very difficult or impossible to measure.

In other words, this clustering of issues is not so much about similar risk outcomes, as it is about similar risk detection methods.

Table 1 Illustration of the format of the BP, as found in the compiled documents published on the website [10]

		Control	Aim
1.1.1	Item name	The control contains the instructions	The Aim provides context and the 'why', often phrased as a risk

The Technical BP focus on the entire product; so not only on the data or the model but encompass the design, integration and overall application of the ML solution to the real world. Its audience is both technical and non-technical stakeholders.

For each of the subjects mentioned above, the items in the Technical BP are sorted into the lifecycle phases (Product Definitions, Exploration, Development, and Production). Each item consists of: a number and a name for easy reference; a *control*, which specifies instructions; and an *aim*, which details the reasons for this control but also provides additional context that will help in resolving any issues that arise during execution (Table 1).

The Organisational BP (OBP) are about how to effectively and structurally support the product teams in an organisation. The end goal is for ML to be embedded across the organisation instead of concentrating in an isolated department. The OBP's components include:

A. **Organisation**—Macro-level organisational structures to ensure oversight for ethical ML.
B. **Product and Model Management**—Responsibilities of Product Owners and team and department managers for a robust support structure.
C. **Resource Management**—To provide (financial resources for) required hardware, software and services, but also to warrant human resources and diversity of skills
D. **Incident Response**—How to respond as an institution to incidents related to, for example, security or safety of ML products
E. **Ethics, Public Interest and Legal**—Ancillary functions to support ML in the organisation as a whole.

Although the contents of the Organizational BP are aimed more at the management, governance, compliance and legal professionals, their application is no less important to the technical professionals in the product teams. It is the anecdotal experience of many of our Foundation members (first-, second- or third-hand) that although the majority of developers really want to create reliable and safe products, they will often be worn down by all the demands placed upon them. This is true for even the most diligent people if they have to deal with too many diverging and time-consuming tasks. These range from:

- red tape and procedural (privacy assessments, risk assessments, data sharing limitations, stakeholder management, etc., etc.) to

- technical subjects that are outside of a typical ML developer's core skillset (data engineering, infrastructure engineering, UX design, security, front-end, etc., etc.), and finally to
- the actual work an ML developer has the proper expertise in and passion for (but by then it's already too late).

In a nutshell, the Organisational BP are there, so that the organisation can provide the correct resources, skills, delegation, ownership of responsibilities, and incentives. This enables every aspect of the ML product development to receive the care it requires for a smooth, efficient and above all safe and responsible conclusion.

4 Conclusion

The Foundation's Best Practices have seen their first release on 19 May 2021 and have been freely accessible since that time for anyone to use (per the creative commons licence) and contribute to through the open source platform.

Although it is the first and only of its kind as far as we are aware, in our opinion the BP align well with recent publications by various regulators [11, 12].

The Foundation's Best Practices are the result of the efforts of a diverse group of volunteers from around the globe and from many professional backgrounds. This diversity is central to the continued validity of the BP and it is therefore the Foundation's mission to increase community interaction as much as possible and incorporate their input in future releases.

References

1. The New York Times (2020) Another arrest, and jail time, due to a bad facial recognition match. https://www.nytimes.com/2020/12/29/technology/facial-recognition-misidentify-jail.html. Accessed 24 Aug 2021
2. The Guardian (2018) Amazon ditched AI recruiting tool that favored men for technical jobs. https://www.theguardian.com/technology/2018/oct/10/amazon-hiring-ai-gender-bias-recruiting-engine. Accessed 24 Aug 2021
3. BBC (2019) Apple's 'sexist' credit card investigated by US regulator. https://www.bbc.com/news/business-50365609. Accessed 24 Aug 2021
4. Artificial Intelligence News (2020) Medical chatbot using OpenAI's GPT-3 told a fake patient to kill themselves. https://artificialintelligence-news.com/2020/10/28/medical-chatbot-openai-gpt3-patient-kill-themselves/. Accessed 24 Aug 2021
5. The Verge (2018) IBM's Watson gave unsafe recommendations for treating cancer. https://www.theverge.com/2018/7/26/17619382/ibms-watson-cancer-ai-healthcare-science. Accessed 24 Aug 2021
6. Partnership on AI (2021) AI incident database. https://incidentdatabase.ai/. Accessed 20 Aug 2021
7. Dao D (2021) Awful AI. https://github.com/daviddao/awful-ai. Accessed 20 Aug 2021
8. Hickok M (2021) AIethicist.org. https://www.aiethicist.org/. Accessed 20 Aug 2021

9. Hagendorff T (2020) The ethics of AI ethics: an evaluation of guidelines. Mind Mach 30:99–120. https://doi.org/10.1007/s11023-020-09517-8
10. The Foundation for Best Practices in Machine Learning. https://www.fbpml.org/
11. European Commission (2021) Proposal for a regulation of the european parliament and of the council laying down harmonised rules on artificial intelligence (artificial intelligence act) and amending certain union legislative acts
12. Information Commissioner's Office (2021) Guidance on AI and data protection. https://ico.org.uk/for-organisations/guide-to-data-protection/key-data-protection-themes/guidance-on-artificial-intelligence-and-data-protection/. Accessed 20 Aug 2021

Recruitment AI Has a Disability Problem: Anticipating and Mitigating Unfair Automated Hiring Decisions

Selin E. Nugent and Susan Scott-Parker

Abstract Artificial Intelligence (AI) technologies have the potential to dramatically impact the lives and life chances of people with disabilities seeking employment and throughout their career progression. While these systems are marketed as highly capable and objective tools for decision making, a growing body of research demonstrates a record of inaccurate results as well as inherent disadvantages for historically marginalised groups. Assessments of fairness in Recruitment AI for people with disabilities have thus far received little attention or have been overlooked. This paper examines the impacts to and concerns of disabled employment seekers using AI systems for recruitment, and discusses recommendations for the steps employers can take to ensure innovation in recruitment is also fair to all users. In doing so, we further the point that making systems fairer for disabled employment seekers ensures systems are fairer for all. disability, artificial intelligence, recruitment, human resources, hiring

1 Introduction

Artificial Intelligence (AI) and similar advanced data analytics systems are increasingly sought-after tools for recruitment used to automate time-consuming, repetitive operational tasks, and expand strategic potential by employers. However, the complexity of engineering that powers the desirable capabilities of these systems, also produces downstream difficulties for organisations to validate the technology for purpose. Uncertain confidence in the decision process and outcomes of technologies involved in recruitment introduces risk in the hiring practices and amplifies the probability of unfair treatment toward job seekers. We posit that without recognition of the

---

S. E. Nugent (✉)
Institute for Ethical AI, Oxford Brookes University, Oxford OX3 0BP, UK
e-mail: snugent@brookes.ac.uk

S. Scott-Parker
Scott-Parker International, London, UK

Business Disability International, London, UK

© The Author(s), under exclusive license to Springer Nature Switzerland AG 2022

M. I. A. Ferreira and M. O. Tokhi (eds.), *Towards Trustworthy Artificial Intelligent Systems*, Intelligent Systems, Control and Automation: Science and Engineering 102,
https://doi.org/10.1007/978-3-031-09823-9_6

limitations of the systems and preventative governance strategies by employer, AI-powered recruitment technologies have the potential to dramatically impact the life chances and economic opportunities of people with disabilities seeking employment.

Commercial AI technologies used in recruitment are marketed as highly capable and objective tools for decision making. In contrast, a growing body of research demonstrates a record of inaccurate results as well as inherent disadvantages across historically marginalised people [1–3]. Assessment of disability fairness in Recruitment AI has thus far received little attention or been overlooked see [4–8].

These findings accompany a landscape of limited regulation and increasing societal pressure for AI and data analytics systems to be designed with fairness, transparency, and validity. This confluence of pressures means that organisations face financial, legal, reputational, operational, and ethical risks for implementing them. While there is already much work being done to address the high-level concerns related to artificial intelligence, bias, and fairness, there will inevitably be more challenges ahead that no one company or industry can solve alone. Minimising these risks requires employers, human and disability rights campaigners, and academic experts to collaborate to develop approaches to validate systems and the design governance strategies to mitigate risks to the job seekers.

Our aim in this paper takes a consumer protections perspective to understanding and mitigating the risks posed by AI recruitment technology. We review the broad range of technological solutions that support the recruitment process, assess the potential and range of impacts to disabled job seekers. We conclude by discussing the options employers having in preparing for procuring on new systems and in evaluating system currently in use.

2 Disability Marginalisation and Employment

As defined by the United Nations Convention on the Rights of Persons with Disabilities (CRPD), "persons with disabilities include those who have long-term physical, mental, intellectual or sensory impairments which in interaction with various barriers may hinder their full and effective participation in society on an equal basis with others."

The definition of disability does not necessarily capture the complexity and heterogeneity disability, which is a key factor that causes complications with AI systems. A disability may be a life-long condition or occur at different life stages or be the result of a major event/change. Disability can have wide-ranging life impacts or be context dependent. Disability may be visible, but most are invisible. Disabilities may include people with hearing, sight and mobility, and dexterity impairments, people with cognitive and intellectual impairments, those with mental health conditions, those with facial disfigurements, those of small stature, and numerous others. Further, individuals may have a combination of multiple disabilities.

Disability is not completely independent of other features of a person's identity and life experience [9–11]. The structural forces and social stigmas that impact people with disability are shared and amplified at the intersection other historically marginalised identities, such those based on as gender, ethnicity, sexuality, and socioeconomic status [12]. Our focus on disability is intended to contribute to a wider discussion of systemic and persistent oppression of marginalized peoples. Recognising and celebrating human diversity is a necessary starting point to design AI systems that fairly and equitably engage with human reality.

People with disability have historically and continue to be regularly disadvantaged in seeking and securing employment. Disabled people experience widespread economic and societal exclusion and are more than twice as likely to be unemployed as able-bodied people [13]. Presently, the sheer scale of the social and economic impacts of the COVID-19 pandemic on employment and employability will undoubtedly further disenfranchise people with disabilities. The current climate of instability makes ensuring fair and equal treatment all the more important, given increasing employment among people with disabilities helps raise people out of poverty, improve their life chances, and is a net cultural and economic benefit [14].

Disability inclusion in the workplace is impacted by number of factors. There is often a qualifications gap between disabled and able-bodied people due to systematic disadvantages in education, training, and previous work experience [15]. Even well-intentioned employers may struggle to recognize how a candidate's previous structural barriers to success impact their present assessment of the candidate's eligibility.

Furthermore, some industries or types of position lack accessibility that can limit employment for people with certain impairments. This maybe due to the employer lacking resources (in the form of advisory programmes) to support the employer in supporting persons with disabilities. Employers may also have negative attitudes and lack confidence or training to support disabled employment seekers [16, 17].

The impact of AI aside, the structural issues affecting people with disabilities in gaining and maintaining employment is a complex and ongoing concern. Therefore layering a complex and opaque system of automated assessment of candidates risk complicating the situation and expanding the risk of harm further.

3 Recruitment AI

As organisations increase in scale and receive larger volumes of job applicants, they are under pressure to balance often competing interests in recruiting and retaining the talented candidates, optimising workflow efficiency and productivity, and managing costs. This means that employers are increasingly turning to automated tools to support the employee's journey from recruitment to retirement.

Artificial Intelligence (AI) has featured prominently in these developments. AI is a subfield of computer science, focused on training computers to perform traditionally human tasks. AI systems are currently available across a wide range of recruitment functions, including: Candidate Sourcing, Engagement Candidate Tracking, CV Screening, Pre-Employment Assessments, and Video Interviewing.

The unifying objective for systems operating across these diverse recruitment functions is that they are designed to distill the vast array of information about applicants down to a few select predictable features for the purpose of making quantifiable and easily comparable decisions. However, when systems need to cope with the reality of human diversity, whether it pertains to disability, ethnicity, gender, and other features, the complexity may be interpreted as abnormality. In this case predictability may come at the expense of the life chances of disabled people who are already faced with systematic disadvantages and unfair discrimination in securing employment.

4 Exclusion by Design and Discriminatory Use

Recruitment AI risks inadvertently, but adversely impacting employment seekers with disabilities via two major routes: biased systems and discriminatory processes.

4.1 Biased Systems

The design of an AI system involves first specifying an objective and then specifying how the system achieves and optimizes achieving that objective. When an objective is not specified adequately, the assessment may lead to unintended consequences in the outcome.

Unwanted biases, or biases that treat people negatively, or adversely due to protected characteristics or other features of their identity within an AI system, raise serious risks of discrimination. It is critical to identify and mitigate these potentially harmful biases.

Unwanted biases relevant to marginalised people, including people with disability, are primarily introduced by historical hiring decisions. Since people with disabilities are already twice as unlikely to be unemployed, they are consequently less likely to be represented in data on past successful employees. These biases may be introduced into systems through two mediums: the algorithmic model and the training data.

The algorithmic model is the mathematical process by which an AI system performs a certain function. Designing this model involves (1) defining the objective or problem the developer determines and (2) selecting the parameters that define the operation of the system at an optimal level [18].

A scenario in which this may occur is if automated CV screener is programmed to predict the best qualified candidate based on an exclusionary parameter, such as having attended a top-tier university. The prestige of an institution may be one factor in a successful employee, but that parameter also disadvantages historically marginalised people, including people with disabilities, who already face systemic barriers to be equally represented in prestigious institutions.

The training data is the preliminary data from which an system learns how to apply the model and produce results in application [18]. The model operates as well as the supplied training data. Bias may be introduced in multiple points along the process before learning begins, from decisions made at data collection to data cleaning to selection for the purpose of training.

Building on the previous scenario, problems may arise with an automated CV screener that was trained on data that did not include the profiles of successful, yet historically marginalised employees. Interacting with information in a CV that the programme has not previously encountered means that the system may be more likely to reject a candidate. This is because these novel or "unusual" features do not fit the prescribed collection of features that is modelled to represent the 'ideal' employee. These novel features may be innocuous, but they may also be indirectly related to the experience of being disadvantaged on the job market.

4.2 Improper Implementation and Use

Even as systems become more technically sound with regard to acknowledging and mitigating bias in design, risks for applicants with disability may be generated and/or amplified by improper use and implementation of the technology.

Most recruiters recognize that no single assessment method is suitable and fair for all applicants [19]. However, the marketed reliability and the ease of automated adaptations of recruitment processes has resulted in many cases where AI tools are being used in isolation of other measures of suitability and human decision makers in the application package. In some organisations, a single product may be the sole gate of entry into employment.

Moreover, AI assessment fails to factor in the likelihood that the employer would make the adjustment post-offer that would determine if a particular disabled candidate was 'right' for the job. For example, a qualified, visually impaired, cybersecurity expert will only be the best candidate if the employer enables them to use specialized software.

Acknowledging and monitoring uncertainty in AI systems is critical to making fair and adequate decisions as sensitive and life changing as whether a person is offered employment or not. The life chances of job seekers precariously intersect with the computational complexities related to disability, the inherent challenges of bias, and the uncertainty around automated decision-making. No system should be expected to–or at least marketed as–working perfectly.

The use of a rigid, standardised recruitment processes that cannot be adequately adjusted to enable candidates with disabilities to compete fairly are inherently discriminatory [20]. Candidates may have the option to request accommodations to these systems—although some developers expect this is the role of the employer to deliver such adjustments. Unless candidates are given explicit assurances that they may request and be provided with equally-evaluated, alternative routes, the employer risks, at best, making disabled users uncomfortable or fearful of interacting with AI and, at worst, discriminating against such candidates. Expecting disabled employment seekers to go through standardised processes is analogous to asking a wheelchair user to take the stairs to the interview room.

5 AI on the Market: The Risks of Discrimination

Recruitment AI encompasses a wide array of technologies functioning at different points in the recruitment process. We outline the broad categories currently in use, detailing the impact potential for people with disabilities. This list is by no means exhaustive, but highlights major technologies used in the candidate sourcing and selection phases of recruitment.

5.1 ATS and CRM Systems

Applicant Tracking Systems (ATS) are platforms where recruiters can conduct each step in the hiring process from posting position openings to collecting applications to screening candidates to evaluation and selection. Candidate Relationship Management (CRM) systems maintain a connection between recruiters and employment seekers so that desirable candidates may be referred to consider future job openings.

We consider these systems together because they share similar impact risks on people with disabilities. They are likely to utilise automated outlier detection tools, such as CAPTCHAs, that when insufficiently trained can flag people with disabilities as not human, or a spammer [4]. People with difficulties related to dexterity or visual impairment are disproportionately affected. The difference between human and non-human may come down to a few seconds delay in response, a minor slip in highlighting the correct answer, or misinterpreting an obscured set of letters.

Further, the skills and qualification gap for disabled people due to systemic inequalities likely disadvantages these candidates when evaluated against the standard person specification as well as historic hiring decisions. These systems are not designed with the flexibility that would take into account that some candidates appear less qualified as a reflection of previous denial of educational and employment opportunities.

5.2 CV/Resume Screeners

CV screening is a major feature of the recruitment innovation powered by AI, addressing the need for processing high application volumes. Automated screeners detect characteristics in the CV content, such as key phrases and words to evaluate employability against criteria for the position. These criteria are determined by either the job description or by evaluating the features of previously successful candidates. They may go further to interpret characteristics of the applicant, such as personality, sentiment, and demographics. Some systems also supplement data in CVs with information about the candidate from public data sources, social media, and information about their previous employers.

Once again, the skills and qualification gap for disabled people due to systemic inequalities is likely to disadvantage these candidates when evaluated against a standard job description as well as historic hiring decisions. These systems are not designed with flexibility that considers some appear less qualified due to systemic lack and denial of education and employment opportunities.

AI screener systems that have not been trained on CV data from users with diverse cognitive and intellectual abilities may have additional challenges with linguistic flexibility. For screeners that analyse personality and emotion from texts, further problems may arise. For example, people with neuro- and cognitive diversity may express emotion in writing in a style previously not encountered by the AI system, resulting in incorrect classifications about their emotional state or personality.

5.3 Conversational Agents

Recruitment conversational agents, or chatbots, are designed to mimic human conversational abilities during the recruitment process. These technologies use natural language processing (NLP) to analyse questions and comments and to respond effectively. Conversational agents are desirable additions to the recruitment process as a means of increasing communication with employment seekers in order to answer frequently asked questions, collect information on candidates, ask screening questions, and schedule interviews or meetings with a human recruiter.

Conversational agent systems have the potential to be helpful in some circumstances where they are designed with accessibility in mind. Agents that augment text with visual illustration (i.e. highlight key words, spelling and grammar check, text suggestion), speech functionality, and dictation tools can enhance accessibility and usability for a wide range of users [21].

However, if not designed and implemented considering disability, conversational agents may also not respond appropriately, or in a hateful manner, and unfairly screen out candidates. Depending on the nature of the agent's function this can at best lead to poor user experience and at worst discriminatory candidate screening.

Conversational agents are often not trained on language data gathered from people with cognitive, intellectual, physical and linguistic diversity or those from neurodiversity groups. Undertrained agents may be unable to correctly interpret spellings or phrases they haven't previously encountered, such as messages from people who have physical difficultly typing or have dyslexia, autism, dysphagia, dyspraxia, ADHD, among numerous others. Moreover, agents that cannot support communications methods beyond writing, such as text-to-speech and dictation, limit or exclude many individuals from participating in communication and being competitive in the recruiting process.

5.4 *Pre-employment Assessments*

A range of candidate aptitude assessments, such as cognitive ability, technical skills, personality, and decision making, are a commonly used to quantitatively measure and compare job applicants for a particular role. Broadly, these tests are aimed at gauging a candidate's ability to think quickly, solve problems, and interpret data.

There is general recognition that these assessments are often not reliable as one-size-fits-all approaches [22]. The generalisability of psychometric tests for people with disabilities, as well as numerous other backgrounds, is unreliable [23]. There is a degree of uncertainty about whether any assessed candidates, never mind those with disabilities, are indeed able to successfully learn and perform the duties of the role or not. Furthermore, many psychometric tests are themselves inaccessible to a wide range of disabled candidates. These assessments must be balanced by other measures in the recruitment process.

Gamified assessments raise additional concerns related to dexterity, vision impairment, and response time. Games often involve tasks that are assessed based on speed of reaction to prompts and precision of responses, which may affect people with motor limitations, who need extra time or assistance to complete dexterity tasks. People with visual impairment may require magnification and colour adjustment and additional time. Furthermore, people with cognitive diversity may require language adjustment and additional time to read prompts [24].

5.5 *AI Interviewing*

AI powered interviewing includes facial analysis tools and speaking conversational agents—aka robot recruiters (refer above to limitations of Chatbots). These tools evaluate employability from the language, tone, and facial expressions of candidates when they are asked an identical set of question in a standardised process. Candidates are assessed based on a variety of facial, linguistic, and non- verbal measures. 'Ideal' measures often are those that most closely align with the same measures from historically successful candidates for any given role.

As with previous examples, systems that are not trained on a diverse range of potentially successful candidates, face challenges in fairly assessing people with facial features, expressions, voice tone, and non-verbal communication that it has not previously encountered [25].

For instance, facial analysis software may inaccurately assess and potentially exclude people with facial disfigurement or paralysis as well as conditions such as Down syndrome, achondroplasia, cleft lip/palate, or other conditions that result in facial differences. Further, people with blindness may not face the camera or make eye contact in a manner acceptable to the system's parameters. Moreover, issues may exacerbated by differences in eye anatomy and dark glasses. People who need captions due to hearing loss, or who lip read may struggle to hear or interpret the questions.

Facial analysis tools that go further to interpret emotion and personality from facial expressions pose alarmingly high risks. Beyond issues of accuracy and algorithmic bias, the fundamental scientific concepts behind personality assessments derived from facial feature measurements, is not supported -and is rooted in pseudoscientific race studies [2]. The implementation of these technologies for recruitment risks legitimising the flawed methodological premise in a way that perpetuates historic disadvantages and exclusion for marginalised peoples.

6 Intervention Recommendations

Designing and implementing Recruitment AI systems that treat persons with disabilities and by that extent, all job seekers fairly requires the engagement of all stakeholders—technology suppliers, purchasers, and users alike. Our aim is to facilitate purchasers in joining the discussion and be prepared with the tools and language needed asks: How do we assess if any given Recruitment AI system is 'safe' for employment seekers with disabilities and others disadvantaged in any labour market?.

There are a number of actions employers can take to procure and implement technological systems that adapt to the values and expectations of their organisation and societal stakeholders toward disabled job seekers. This process begins by introducing disability into governance plans across departments. Involving multiple stakeholders in asking disability-conscious questions of technology suppliers increases the likelihood of identifying and mitigating risks before the system begins making critical decisions about job seekers.

6.1 Executive Level Stakeholders

Starting at the executive level, addressing disability inclusion in recruitment may be assessed as part of wider strategic and corporate vision considerations. Suppliers can be assessed for suitability based on if

i. the recruiting system supports organizational goals to increase diversity and representation.
ii. the recruiting system follows any existing policies with regard to the ethical and responsible development and implementation of artificial intelligence?
iii. the supplier actively engages in learning how to adapt to match organisational values and needs as a business and stakeholders?

iv. which employees should be involved in the governance process which determines how to investigate, procure, apply and monitor recruitment systems so that they do not adversely impact disadvantaged job seekers?

6.2 Human Resources Stakeholders

Recruitment AI supports tasks that would otherwise be undertaken by HR or recruitment professionals. These individuals are well-suited to evaluate if the automated process mirrors protocols that would be expected of human practitioners. Considerations for disability-inclusive suitability of the system include:

i. the benefits and risks of the system for disabled and other disadvantaged employment seekers
ii. that the supplier has a shared understanding of inclusivity and fairness—with specific reference to eliminating the root causes of disability related discrimination—designed into the system.
iii. if implementing the system requires alternative evaluation routes to enable people with different impairments to be recruited on the basis of individual capability and potential.
iv. if recruitment system enables candidates to readily request adjustments, in a non-stigmatising manner, at every stage of the process.

6.3 Procurement Stakeholders

Procurement is the first entry point for recruitment systems into an organisation. Procurement professionals are in a position assess disability inclusivity of a recruitment system as part of the risk management process. Considerations may include:

i. if this supplier can demonstrate that their product is safe for disabled and other disadvantaged employment seekers before you purchase
ii. How has the supplier actively involved people with disabilities to test and validate its products?
iii. Was a shared understanding of inclusivity and fairness—with specific reference to eliminating the root causes of disability related discrimination—designed into this technology?
iv. Do contractually defined performance standards require the supplier to track the experience of job seekers with disabilities—particularly those who have requested disability related adjustments?
v. Can they evidence that they have actively consulted and involved persons with disabilities as expert advisors and potential users in their product development life cycle?

6.4 *Information Technology Stakeholders*

Finally IT professionals will also need to be prepared to consider the role of disability in the technical functionality of recruitment systems. They will be in a position to evaluate:

i. if they have been provided with the appropriate explainability and interpretability resources to assess outputs and impacts on employment seekers' disabilities.
ii. if the relevant, quality data exist to support this technology in performing effectively for persons with disabilities.
iii. if the appropriate oversight mechanisms are in place to evaluate the performance of the system and can the system withstand scrutiny by disabled employment seekers.
iv. if the supplier demonstrate how the processes will adapt so as to ensure equal opportunities for disabled employment seekers.

Acknowledgements We dedicate this chapter in the memory of James Partridge (Face Equality International) who contributed valuable insights to this research and whose advocacy transformed attitudes toward people with facial disfigurements. We appreciate the contributions and insights provided by Julien Burnett, Nigel Crook, Paul Jackson, and Rebecca Raper.

References

1. Broussard M (2018) Artificial unintelligence: how computers misunderstand the world. MIT Press, Cambridge
2. O'Neil C (2017) Weapons of math destruction: how big data increases inequality and threatens democracy. Penguin Random House, New York
3. Noble S (2018) Algorithms of oppression: how search engines reinforce racism. New York University Press, New York
4. Guo A, Kamar E, Vaughan JW, Wallach H, Morris MR (2019) Toward fairness in AI for people with disabilities: a research roadmap. arXiv preprint arXiv:1907.02227
5. Petrick ER (2015) Making computers accessible: disability rights and digital technology. Johns Hopkins University Press, Baltimore
6. Trewin S (2018) AI fairness for people with disabilities: point of view. arXiv preprint arXiv: 1811.10670
7. Trewin S, Basson S, Muller M, Branham S, Treviranus J, Gruen D, Hebert D, Lyckowski N, Manser E (2019) Considerations for AI fairness for people with disabilities. AI Matters 5(3):40–63
8. Whittaker M, Alper M, Bennett CL, Hendren S, Kaziunas L, Mills M, West M (2019) Disability, bias, and AI. AI Now Institute, November
9. Collins PH, Bilge S (2020) Intersectionality. Wiley and Sons
10. Parker AM (2015) Intersecting histories of gender, race, and disability. J Women's Hist 27:1
11. Samuels E (2016) Fantasies of identification: disability, gender. New York University Press, New York, Race
12. Frederick A, Shifrer D (2019) Race and disability: from analogy to intersectionality. Sociol Race Ethnicity 5(2):200–214

13. Office for National Statistics, Disability and Employment, UK (2019). https://www.ons.gov.uk/peoplepopulationandcommunity/healthandsocialcare/disability/bulletins/disabilityandemploymentuk/2019
14. Schur L (2002) The difference a job makes: the effects of employment among people with disabilities. J Econ Issues 36(2):339–347
15. Sayce L (2011) Getting in, staying in and getting on: disability employment support fit for the future, vol 8081. The Stationery Office
16. Lindsay S, Leck J, Shen W, Cagliostro E, Stinson J (2019) A framework for developing employer's disability confidence. Equal Divers Inclus Int J (2019)
17. Suter R, Scott-Parker S, Zadek S (2007) Realising potential: disability confidence builds better business. Employers' Forum on Disability
18. Russell S, Norvig P (2003) Artificial intelligence: a modern approach, 2nd edn. Pearson Education
19. Davis L (2005) Disabilities in the workplace: recruitment, accommodation, and retention. AAOHN J 53(7):306–312
20. Hamraie A (2017) Building access: universal design and the politics of disability. University of Minnesota Press, Minneapolis
21. Følstad A, Brandtzæg PB, Feltwell T, Law EL, Tscheligi M, Luger EA (2018) SIG: chatbots for social good. In: Extended abstracts of the 2018 CHI conference on human factors in computing systems, pp 1–4
22. Edenborough R (2005) Assessment methods in recruitment, selection and performance: a manager's guide to psychometric testing, interviews and assessment centres. Kogan Page Publishers
23. Smith K, Abrams SS (2019) Gamification and accessibility. Int J Inform Learn Technol
24. Cook DA, Beckman TJ (2006) Current concepts in validity and reliability for psychometric instruments: theory and application. Am J Med 119(2):166–167
25. Leslie D (2020) Understanding bias in facial recognition technologies. arXiv preprint arXiv:2010.07023

Developing and Evaluating Complex Interventions: The Case of Robotic Systems in Cognitive Rehabilitation Therapy

Maria Isabel Aldinhas Ferreira

Abstract The use of robots in cognitive rehabilitation has been going on, at an experimental level, in the last decades, comprehending a broad scope of applications ranging from those aiming at the stimulation of patients with dementia to others addressing children in the autism spectrum. Though the results provided by the multiple projects developed in this domain apparently point to the eventual beneficial character of this intervention, its validation and formal recognition in clinical terms still lacks, keeping these experiments punctual and inconsequent in medical terms, being the impact of its developments limited to the robotics R&D. In the present paper, we briefly address this issue claiming that the formal clinical recognition of the possible beneficial role of artificial intelligent systems in cognitive rehabilitation therapy is only possible if/when the robotic projects dedicated to this theme are designed and developed as Complex Interventions, i.e., interventions that stand on a lattice of interdependencies and co-effects. According to the Medical Research Council, Complex Interventions are widely used in the health service, in public health practice, and in areas of social policy such as education, transport and housing that have important health consequences. The present paper reviews the guidelines and recommendations defined both by the "Framework for Development and Evaluation of Randomized Control Trials (RCTs) for Complex Interventions to Improve Health", April 2000, and its updated 2019 version "Developing and evaluating complex interventions". Both are intended to help researchers with the design of the intervention, choosing the appropriate methods, identifying the constraints on evaluation, presenting the evidence in the light of methodological and practical constraints. The present paper adopts an end-user centered stance in which the physical and psychological well-being of the patients that are targeted by these interventions, namely those that volunteer to participate in the trials, as well as that of their families, is viewed as ethically prior. According to this assumption, a fundamental

This was the theme of a communication presented by the author at the International Conference on Robot Ethics and Standards-ICRES 2020, 27–28th September 2020, Taiwan (online participation).

M. I. A. Ferreira (✉)
Centre of Philosophy of the University of Lisbon, Lisbon, Portugal
e-mail: isabelferreira@letras.ulisboa.pt

Institute for Robotics and Systems, ISR/IST, University of Lisbon, Lisbon, Portugal

© The Author(s), under exclusive license to Springer Nature Switzerland AG 2022
M. I. A. Ferreira and M. O. Tokhi (eds.), *Towards Trustworthy Artificial Intelligent Systems*, Intelligent Systems, Control and Automation: Science and Engineering 102,
https://doi.org/10.1007/978-3-031-09823-9_7

role is given to the monitoring of this well-being and its assessment throughout the intervention process and the monitoring and assessment of the evolution of their health condition before, during the trials and after the experiment, in a long-run mode.

1 Introduction

Trust is fundamental in healthcare and is the grounding stone of all interactions that are defined within the network of relations responsible for the physical and mental health care of citizens and for their well-being.[1,2,3] It is first substantiated in the relation that the patient holds with their physician[4] and with the other clinical staff, it is also present in the patient's acceptance of a new clinical procedure or a new drug. We have had the chance to observe for example, how trust judgements determined the acceptance or complete rejection of the Covid-19 vaccines by entire populations.

The volunteering to participate in typical clinical trials or those that relate the testing on patients of technologies that are thought to be beneficial, as it is, for instance, the case of exoskeletons in mobility impairments or the use of social assistive robots in cognitive therapy rehabilitation, relies on a trust relationship between parts. Building trust when designing and running a trial is fundamental. As [2] points out, trial professionals and those working with them sometimes think about trust in terms of intellectual credibility, but there are additional dimensions of trust, including reliability (or consistency) and perceptions of shared interest.

Perception of a shared interest is translated in the awareness of the convergence of the distinct motivations that drive the two main players within the trial: on one hand the researcher conducting the trial and their scientific/technological drivers, on the other hand those of the one volunteering to participate in this trial. Awareness of the motivations and expectations of those volunteering is essential especially when this motivation relates to their own physical condition or the health condition of a loved one. Equally important is the awareness of expectations relative to the potentially beneficial character of what is being tested on the improvement of a physical or mental condition.

When running a trial, the patient's health status and its evolution must be accompanied not just during the trial but also on the long run. This post trial phase, this essential long-term monitoring, is generally absent from the trials run by robotics R&D. This absence compromises the clinical credibility of the results obtained and

---

[1] https://mrc.ukri.org/documents/pdf/complex-interventions-guidance/.

[2] https://mrc.ukri.org/documents/pdf/rcts-for-complex-interventions-to-improve-health/.

[3] https://mrc.ukri.org/documents/pdf/complex-interventions-guidance/.

[4] According to Johanna [1], the patients' trust in their health care professionals is central to clinical practice and has been identified as the foundation for effective treatments and fundamental for patient-centered care.

the reliance by health professionals on the benefits the technology may bring to patients.

R&D robotic trials generally take place within the framework of a funded project and according to the project's timeline and defined milestones. The dynamics of these projects is never patient-centered, though the volunteers are a part the project could not live without. During these projects, robotic technology aiming to contribute to solve or ameliorate a particular health condition is developed, its safety tested in the lab and then its potential benefit tested in trials with volunteer patients. Typically, by the end of the project's timeline, results are presented, and the project is considered successfully accomplished. However, besides the clinical inconsistency that is due to a general hyper fragmented modus operandi—though individual projects partially feed on their predecessors technological achievements, they in fact, in clinical terms, seem always to be restarting from the zero—these trials raise a fundamental ethical issue: how have they responded to the volunteers' motivation and their legitimate expectations? The answer is, in most cases, very unsatisfactory. In fact, once the trial data is collected and presented the project's goals are considered accomplished. Both researchers and technology withdraw, and volunteers and their families go back to" ground zero".

In this paper, we advocate that to have their value clinically recognized and in order to correctly address the ethical issues concerning of volunteer participants, R&D robotic trials should be patient-centred and follow the procedures close to those established for Complex Interventions.

2 Complex Interventions: Definition and Scope

Complex interventions are widely used in the health service, in public health practice, and in areas of social policy such as education, transport and housing that have important health consequences. According to the Framework for Development and Evaluation of Randomized Control Trials (RCTs) for Complex Interventions to Improve Health [3], complex interventions are built up from a number of components, which may act both independently and inter-dependently. These components are comprehended by the variables: Patient, Context and Therapy and usually include behaviors, parameters for observing these behaviors (e.g. frequency, timing), and methods of organizing and delivering those behaviors (Fig. 1).

The semantic scope of the term [Complex Intervention] is consequently broad comprehending processes of varied nature and degree of complexity. As pointed out in Developing and Evaluating Complex Interventions (2019), at the simplest end of the spectrum we can situate the randomized control trials of a drug versus a placebo and at the most complex the assessment of a specialized care unit versus traditional care (Fig 2).

Though each intervention demands fine attuning according to its specific context, variables and parameters involved, there are general parallels in the sequence of

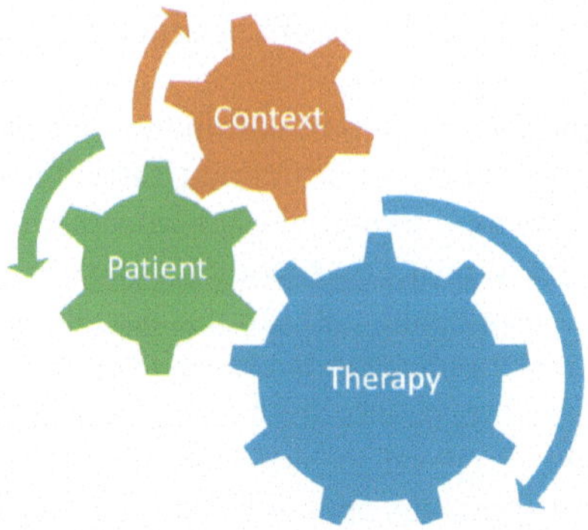

Fig. 1 Complex intervention variables: the patient, the context and the therapy

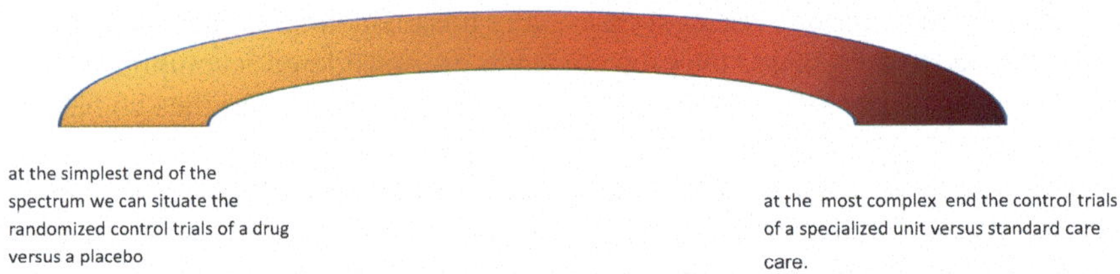

Fig. 2 The semantic scope of the term complex interventions is broad comprehending processes of varied nature and degree of complexity

steps usually required in the evaluation of drugs from initial pre-clinical research to post-marketing surveillance and the complex intervention packaging.

According to this perspective, Developing and Evaluating Complex Interventions (2019), identifies the following phases in a Complex Intervention:

- 'Pre-Clinical' or theoretical
- Phase I or modelling
- Phase II or exploratory trial
- Phase III or main trial
- Phase IV or long term surveillance (Fig. 3).

The 'Pre-Clinical' or theoretical phase: Refers to the gathering of the formal bodies of evidence where they exist or, when these do not exist yet, the wisdom grounded on empirical evidence.

Phase I or modeling involves delineating an intervention's components and how they inter-relate and how active components of a complex package may relate to either surrogate or final outcomes.

When you undertake cognitive behavioral therapy, what exactly is or are the "active ingredient(s)"? Does success depend on the personality of the therapist? The personality, health status, social status, or other characteristic of the patient? The

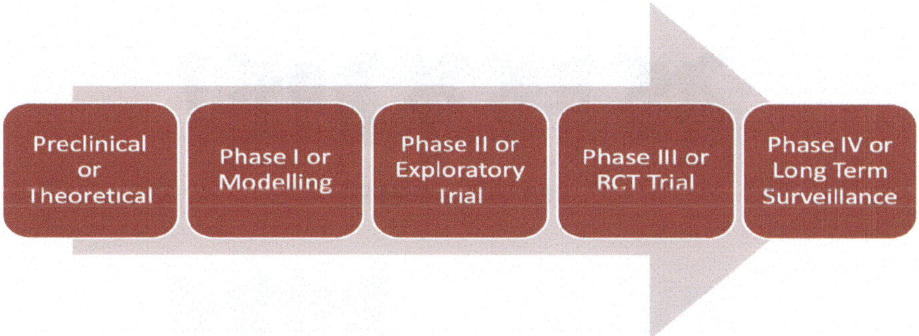

Fig. 3 The phases in a complex intervention

content of the therapy? The way it is delivered? The frequency of contact? The location of contact? The duration and the timing? What other components count?

In this phase, simulations (which include computer modeling) can help explore questions such as "If I change this factor, what happens to all those factors?". Simulations can be used both to test assumptions and to provide important information regarding trial design needs.

Phase II or Exploratory trial: In Phase II, all the evidence gathered thus far is put to the test. A key question in evaluating complex interventions is how the intervention works, in other words, what are the active ingredients within the intervention and how are they exerting their effect.

Only by addressing this kind of question one can build a cumulative understanding of causal mechanisms, design more effective interventions and apply them appropriately across group and setting.

In Phase II the appropriate control group is identified, outcome measures, estimates of recruitment for a main trial, and other requirements. This phase also permits to test alternative forms of an intervention. The exploratory trial also provides the opportunity to "test-drive" the assumptions and strategies which will form the Phase III design; methods of recruitment, randomization, follow-up can all be examined at this stage.

Phase III or main trial is the corner stone in the evaluation of a complex intervention. It is the main randomized controlled trial (RCT) to evaluate a complex intervention. It requires the selection of outcome measures most appropriate to the evaluation of a specific complex intervention and requires attention to adequate randomization and blinding (where feasible), appropriate outcomes measures, informed consent of participants and other standard features of well-designed trials (Fig. 4).

The volunteers constituting the Experimental Arm test the new drug or therapy, the ones in the Control Arm follow the usual therapy or a placebo.

Phase IV or long-term surveillance Most complex interventions require a long-term monitoring and assessment and the commitment from the previous participants and their families.

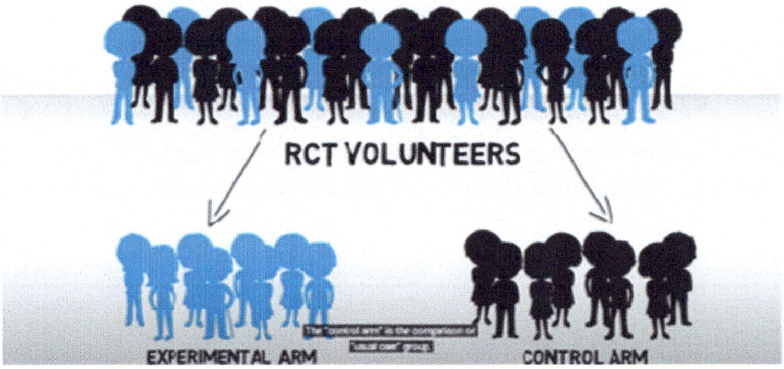

Fig. 4 RCT volunteers: experimental arm and control arm

The fundamental final step in the evaluation of a complex intervention is this long-term monitoring, this observational study to establish long-term term results. A fundamental key question in this evaluation is the identification of its practical enduring effectiveness, to understand the whole range of subsequent beneficial effects and how this effectiveness might have varied among recipients and if it has what were the causes. As important is the identification of emergence of undesirable side effects.

3 Robotics in Cognitive Rehabilitation Therapy (CRT)

In the present paper, the expression Cognitive Rehabilitation Therapy (CRT) designates any intervention based on behavioral education leading to the development of cognitive processes (such as attention, working memory, executive functions (EFs), social recognition, and metacognition) with the aim of generalization and sustainability. Its goal is the improvement of cognitive deficits in a broad spectrum of mental conditions including: psychiatric conditions as schizophrenia, attention deficit/hyperactivity disorder (ADHD), brain damage, and ASD, cf., Atigh and Alizadeh Zarei [4].

Autism spectrum disorder (ASD) is a neurodevelopmental disorder characterized by cognitive and neurobiological deficits based on social and non-social information processing defects, [5] causing a pattern of stereotypical abnormal repetitive behaviors as well as significant deficits in communication. According to the WHO, it is estimated that worldwide one in 160 children has an ASD.[5] This estimate represents an average figure, and reported prevalence varies substantially, however epidemiological studies conducted over the past 50 years show the prevalence of ASD appears to be increasing globally. There are many possible explanations for this

[5] https://www.who.int/news-room/fact-sheets/detail/autism-spectrum-disorders.

apparent increase, including improved awareness, expansion of diagnostic criteria, better diagnostic tools and improved reporting.

Since Dautenhan's first seminal works[6] on this topic, Dautenhan et al. [6], Robins et al. [7], highlighting the potential of robots for children autism therapy, robotic projects targeting ASDs have been proliferating as one of the first application domains in the field of socially assistive robotics.[7] Robot-assisted therapy (RAT) also designated more recently as Robot-Enhanced therapy, Hoang et al. [8], has been put into practice as a tentative form of cognitive rehabilitation therapy to improve the performance of ASD children, [9], Sartorato et al. [10]. And even fiction has, sometimes, reflected the high expectations the deployment of robotic technology in this field has been raising.[8]

However, as [11] and [12] point out, the extensive review works on the typology, scope and methods followed in most of robotics ADS projects has showed that the focus of much of this research is on robot functional development and robot methodology rather than on its clinical effectiveness/efficacy, not providing sufficient details for clinical inquiry and eventual clinical assessment and trial validation. According to those authors, most of this research is based on single case studies and simple case experiments. Few studies make statistical comparisons between groups or between conditions, and none tested the independent contribution of the robot to the clinical application. Also, even with the longitudinal design of some of the studies, the experiments did not necessarily show ability to generalize skills over time, place, or context.

The fact that most of these projects run on a limited time span consequently makes them:

- Punctual,
- Incomplete in what concerns their efficiency
- Non-conclusive.

Consequently:

- Clinically unreliable.

More recently projects as INSIDE (2018) or DREAM (2018) have attempted to incorporate a real clinically-oriented perspective on the way the whole project is designed and structured and on the role of RCT and the data derived from it, coming closer to what is expected from an intervention assumed within the Complex Intervention Framework. However they apparently dramatically lack the long-term monitoring [13].

Abundant rigorous clinical data resulting from the implementation of complex intervention processes, comparing the benefits of artificial intelligent systems in

---

[6] AuRoRA Project http://aurora.herts.ac.uk/.

[7] SoftBank Robotics. Accessed on: Mar. 20, 2019. [Online]. Available: https://www.softbankrobotics.com/emea/en/nao.

[8] When Dinosaurs Ruled the Earth, a short film about autism, robots and dinosaurs. J Autism Dev Disord. 2000 Jun;30(3):205–223.

Cognitive Rehabilitation Therapy with standard therapy is needed to establish its efficacy. Plus, without the observation studies resulting from the monitoring and assessment of phase IV, in the long term no therapy can be clinically validated.

In our opinion any robotic/AI technology after having proved in the lab to be completely safe and reliable, must be tested with end-users in real life scenarios and assessed according to a transparent and well-defined metrics that measures not just the functional performance of the system but its ethical and social adequacy. We advocate that in the particular case of artificial systems targeting the health domain, namely those intended to act as therapeutic tools, RCTs with real patients, in real scenarios, have to be designed following the general framework established for Complex Interventions. In this line, it is needed (i) a rigorous characterization of the patient, the identification of the main contextual factors that will play a role in the process (either as constraints or as positive drivers) and that will determine the choice of a particular therapy and how this therapy should be "dosed" and applied to that particular individual in that specific context; (ii) a robot–enhanced therapy targeting ADS, has to follow Randomized Control Trials in order to gather rigorous clinical results that compare the eventual efficacy/benefits of an innovative therapy versus the standard one. This is the first step to identify the therapeutic and clinical potential of a new form of intervention; (iii) the long-term monitoring of patients is essential to get its clinical validation and in the long run its universal medical/clinical approval.

Only the long-term monitoring will provide the fundamental data for that validation:

(a) identifying the duration of the therapy's effects: Are the beneficial features registered permanent? Are they temporary? Does the therapy need to be reinitiated? What's its frequency? Under which conditions is it necessary to keep this kind of therapy on the long run? If yes, what should its frequency be? In what "dosage"? How can this be achieved?—at home? at a care unit? Who will be responsible for this therapeutic updating? How will it be monitored?
(b) identifying eventual long-term negative side-effects or impacts.
(c) contrasting the data provided by this long-term monitoring with the monitoring of similar congeners in different locations/regions to identify potential variability.

We also believe that only within the Complex Interventions Framework that places the patient and their characteristics in the front stage of the process, can the delicate ethical issues that emerge from the application of this type of technologies be properly tackled.

4 Conclusions

Intelligent technologies have progressively been introduced in health care and health systems, either as advanced forms of therapy, as chirurgical tools, as diagnosis facilitators, or just as communication enhancers, promoting alternative forms of interaction between patients and health support units. However, most of these technologies are far from mature and their positive or the eventual negative impacts on individuals physical and mental health and on society is far from being consistently addressed. An objective coherent evaluation of their performance and eventual benefits on individual and public health is consequently needed so that these tools are either clinically validated, upgraded, or dismissed. Taking the example of the research going on in Robot Enhanced Therapy, we wanted to call the attention of R&D to the need of following a Complex Intervention Design whenever developing innovative robotics/AI approaches to treatment, diagnosis or just interaction in health care. Only within this framework it will be possible to eventually validate innovative therapies, procedures, and tools. Also, this framework will make it easier to be aware of all the circumstances involved and consequently identify the potential emergent ethical issues. The sensitiveness of this particular area was already referred by [14, 15].

Within the very complex ethical framework that emerges from the crisscrossing of both health care ethics and the ethics of robotic and AI driven technologies, we would like to finish calling a special attention to the very delicate position of Robotics/AI R&D when testing innovative tools.

In this context, the main questions are:

(i) What is the value of a research that is not clinically guided and whose results are not clinically validated on the long run?
(ii) Is it ethical to "use" patients to test therapeutic solutions without having a real concern for their own evolution and well-being, outside the framework of a research project?
(iii) Is it ethical that patients after having eventually experienced some progresses within the timeline of a project are not able to continue their interaction with these tools, because the project has finished and those tools have to be removed?
(iv) Is it ethical to raise expectations of a better life and then let these expectations drop?

Though we are aware of the legal framework that distinguishes Clinical Research from Medical Treatment[9] and that defines what the participants in Randomized Controlled Trials are legally supposed to expect or not from their participation in a trial, we would like to highlight that, generally, the main motivation of volunteers and their families is to contribute to the development of scientific knowledge that will ultimately, hopefully, ameliorate their health condition. In this sense, the respect for the dignity and well-being of the participant, that has to be present throughout

[9] https://alea-research.com/clinical-research-vs-medical-treatment/#:~:text=While%20there%20are%20a%20lot,help%20patients%20in%20the%20future.

the intervention, assumes a particular relevant role in Phase IV by guaranteeing that participants will not be left "on their own" and back to the same standard therapies even in the case a new one has proved to work better.

References

1. Birkhäuer J, Gaab J, Kossowsky J, Hasler S, Krummenacher P, Werner C, Gerger H. (2017) Trust in the health care professional and health outcome: a meta-analysis. PLoS One. 12(2):e0170988. https://doi.org/10.1371/journal.pone.0170988. https://www.ncbi.nlm.nih.gov/pmc/articles/PMC5295692/
2. Southwell B (2021) A future of trusted clinical trials: communication strategies to encourage trust and transparency. https://www.healthaffairs.org/do/. https://doi.org/10.1377/forefront.20210503.292254
3. A Framework for Development and Evaluation of RCTs for Complex Interventions to Improve Health. Medical Research Council (2000). https://mrc.ukri.org/documents/pdf/rcts-for-complex-interventions-to-improve-health/
4. Atigh A, Alizadeh Zarei M (2019) The effect of cognitive rehabilitation therapy (CRT) on the executive functions of children with autism spectrum disorder (ASD). Chronic Dis J 137–147. ISSN eISSN: 2345–2226. Available at: http://cdjournal.muk.ac.ir/index.php/cdj/article/view/408/181. Date accessed: 20 Sep. 2020. https://doi.org/10.22122/cdj.v7i3.408
5. Lord C, Risi S, Lambrecht L, Cook EH Jr, Leventhal BL, DiLavore PC, Pickles A, Rutter (2000) The autism diagnostic observation schedule—generic: A standard measure of social and communication deficits associated with the spectrum of autism. J Autism Develop Disorders 30(3):205–223
6. Dautenhan K et al (2004) Towards interactive robots in autism therapy. Pragmatics Cognition 12:1–35. ISSN 0929–0907/E-ISSN 1569–9943_©John Benjamins Publishing Company
7. Robins B, Dautenhahn K, Billard A (2005) Robotic assistants in therapy and education of children with autism: can a small humanoid robot help encourage social interaction skills? Univ Access Inform Soc 4(2):105–120
8. Hoang C, Esteban P, Bartlett M, Baxter P, Belpaeme T, Billing E, Cai H, Coeckelbergh M, Costescu C, David D, De Beir A, Hernandez D, Kennedy J, Liu H, Matu S, Mazel A, Pandey A, Richardson K, Senft E, Thill S, Van de Perre G, Vanderborght B, Vernon D (2019) Inside Project. https://welcome.isr.tecnico.ulisboa.pt/inside-project/
9. Huijnen CA, Lexis MA, Jansens R, de Witte LP (2016) Mapping robots to therapy and educational objectives for children with autism spectrum disorder. J Autism Dev Disorders 46(6):2100–2114
10. Sartorato F, Przybylowski L, Sarko DK (2017) Improving therapeutic outcomes in autism spectrum disorders: enhancing social communication and sensory processing through the use of interactive robots. J Psychiatric Res. 90:1–11
11. Diehl JJ, Schmitt LM, Villano M, Crowell CR (2012) The clinical use of robots for individuals with autism spectrum disorders: a critical review. Res Autism Spectr Disorders 6(1):249–262
12. Pennisi P, Tonacci A, Tartarisco G, Billeci L, Ruta L, Gangemi S, Pioggia G (2016) Autism and social robotics: a systematic review. https://doi.org/10.1002/aur.1527
13. Melo FS, Sardinha A, Belo D, Couto M, Faria M, Farias A, Gambôa H, Jesus C, Kinarullathil M, Lima P, Luz L, Mateus A, Melo I, Moreno P, Osório D, Paiva A, Pimentel J, Rodrigues J, Sequeira P, Solera-Ureña R, Vasco M, Veloso M, Ventura R (2019) Project INSIDE: towards autonomous semi-unstructured human–robot social interaction in autism therapy, Artif Intell Med 96:198–216

14. Wakunuma K, Yu H, Zhou X, Ziemke T (2019) Enhanced therapy-developing and validating a supervised autonomous robotic system for autism spectrum disorders therapy. Available at https://coeckelbergh.files.wordpress.com/2019/04/2019_04_05-coeckelbergh-robot-enhanced-therapy-ieee.pdf
15. Veruggio G, Operto F (2008) Roboethics: social and ethical implications of robotics. In: Siciliano B, Khatib O (eds) Springer handbook of robotics. Springer-Verlag, Berlin, pp 1499–1524
16. DREAM, Clinical Trials (2018) The efficacy of robot-enhanced therapy for children with autism spectrum disorders (DREAM-RCT). Available: https://clinicaltrials.gov/ct2/show/NCT03323931

COVID-19 Contact Tracing Applications in Portugal: Effectiveness and Privacy Issues

Arlindo L. Oliveira

Abstract Portugal developed and adopted a contact tracing smartphone application, in order to help in the containment of the spreading of the COVID-19 pandemic. The particular technology used was adopted in order to guaranteed privacy and security but it did not prove particularly effective, since it was downloaded by only a fraction of the population and very lightly used to report cases of contacts between infected and non-infected persons. These issues are addressed and discussed in the present text.

1 Introduction

The COVID-19 pandemic, which started in Wuhan, China, and rapidly spread to all the continents and countries, was the first major pandemic to take place in the digital age. The two most recent major flu outbreaks, the 1957–1958 Asian Flu [3] and the 1968–1969 Hong Kong flu [7] pandemics, which had comparable impacts in terms of number of infected persons and number of deaths, when adjusted by world population, took place at a time when digital technologies were in their infancy. The 2020 COVID-19 pandemic, on the other hand, took place in the digital age, raising the possibility that digital technologies could be effectively used to control the spread of the disease.

For these reasons, but also because today's society is much more adverse to risk than the society of the mid twentieth century, the COVID-19 pandemic generated an enormous amount of media attention and unprecedented reactions from governments, society and the population in general. Complete lockdowns took place in a significant fraction of European and Asian countries, unprecedented virus suppression measures have been adopted by almost all the world countries and billions of people faced enormous economic challenges caused by the anti-COVID measures adopted by governments. For these reasons, there was a strong pressure on the engineering and data science community to develop methods to model the pandemic, trace the infected

---

A. L. Oliveira (✉)
INESC-ID/Instituto Superior Técnico, Lisbon, Portugal
e-mail: arlindo.oliveira@tecnico.ulisboa.pt

© The Author(s), under exclusive license to Springer Nature Switzerland AG 2022
M. I. A. Ferreira and M. O. Tokhi (eds.), *Towards Trustworthy Artificial Intelligent Systems*, Intelligent Systems, Control and Automation: Science and Engineering 102,
https://doi.org/10.1007/978-3-031-09823-9_8

people and stop the spread of the SARS-CoV-2 virus. These efforts were developed in parallel with all the work done by the life sciences scientific community to treat COVID-19 patients, understand the biological basis of the disease, and create new vaccines and treatments.

The extensive availability of digital mobile devices, internet-enabled smartphones, raised the interesting possibility of using contact tracing applications to help contain the spread of the disease. Contact tracing is potentially useful in the containment of COVID-19, a disease that propagates mainly though physically close contacts between infected and non-infected persons. However, manual contact tracing of a pandemic of these proportions requires a significant amount of resources, advance planning and organized procedures. None of these conditions were present in Portugal in the Winter of 2020, when the pandemic hit the country in full force. For these reasons, the Portuguese government decided to pursue a contact tracing strategy based on the utilization of digital technologies. However, although a contact tracing application was developed and deployed nationwide in September of 2020, it did not play a significant role in the control and containment of the pandemic.

2 COVID-19 Smartphone-Based Contact Tracing in Portugal

A large number of contact tracing applications have been developed and implemented worldwide [5]. In Portugal, a number of initiatives led to the consideration of three different technologies for contact tracing applications. These technologies adopted either a centralized or de-centralized strategy, leading to different tradeoffs of effectiveness, privacy risks and usability.

The first possibility considered involved the development of an application that used the global positioning system (GPS) information available in many smartphones to create a global person-to-person interaction graph derived from the position history determined by the GPS positioning system and registered by the smartphones. The second possibility was to make use of the location information available to mobile phone operators in order to track the locations of all, or a fraction, of the users of mobile phone service, to model and track the spread of the pandemic and, possibly, to identify specific episodes of high transmission risk. The third possibility was to use the Bluetooth low energy (BLE) beacon technology to create a distributed graph of person-to-person interactions that could be used to identify a significant fraction of the contacts between infected and non-infected people. The different approaches involved different privacy/effectiveness tradeoffs, which were considered in the choice of the final solution and are detailed in the next section.

3 Contact Tracing Technologies

The first solution considered required the creation of a centralized database, which stored the history of the location of every person, derived from the GPS information obtained and registered by each smartphone. This database could then be processed to identify all episodes of proximity between different users of the application and to trace the contacts of infected persons. These contacts could then be used to warn possibly infected persons and lead them to take preventive measures or undergo testing. Although the technology to develop such an application in the short-term was available, the need for a centralized database that contained all the information about the location of a significant fraction of the population raised obvious concerns with privacy and the approach was never pursued, not even as a pilot project.

The second possible solution was based on the fact that every mobile phone operator keeps an history of the approximate location of every user. In fact, the operation of the mobile phone network requires the system to identify and store the antennas that are being used by each equipment, at each time, and this information is usually stored for a relatively long period, since it useful for planning and reporting purposes. Due to privacy and security considerations, this information is kept locked and accessible only to a small number of duly authorized persons, but could in principle be unlocked and made available. However this information is difficult to use for tracing purposes, since the precision of the location information is not very high. Depending on the density of the antennas in a given region, the location of the phone user can be determined with only a limited precision, much smaller than the one obtainable by a GPS system. Without additional processing, the specific location of a user can be determined within a range that varies between a few hundred meters and several kilometers. Algorithms that post-process this information, taking into account the signal strength received by different antennas may increase the precision of the location, but the results are still too imprecise to determine if two persons have been in close range of each other, corresponding to a potential risky contact. Therefore, this approach could only be used to identify general transmission trends between regions, or regions of high risk of transmission, and did not pose significant threats to privacy. The location data could also be used to generate mobility matrices that describe how many people moved between different cities or counties in a given period, from hours to days. This could raise a potential privacy problem since, in some particular cases, the numbers of people traveling between two locations are so low that they could compromise the privacy of specific persons. However, the mobile phone operators avoid this threat to privacy by reporting only inter-city mobility numbers when they exceeded some fixed threshold, a procedure that is effective in preserving anonymity. The modeling experiments that were carried out to test the usefulness of this information in the modeling of the territorial spread of the pandemic did not contribute to advance the state of the art [2], as of the time of this writing, and will not be reported here.

The third solution, based on BLE beacons [4], did not require the construction of a centralized database and provided stronger warranties of privacy than the first option.

INESC-TEC, an associated laboratory belonging to the INESC group, developed an application [6]based on the DP-3T framework [8], which was developed explicitly to limit privacy risks caused by COVID-19 tracing applications. The DP-3T framework limits the risk to privacy by storing the information about which phones are in the proximity of each user in that user phone and not in a centralized database. In the absence of a centralized database it is, in principle, unfeasible to an attacker the use of this information to track the behavior of any individual user. DP-3T is based on the idea that each individual phone can keep an encoded key of all the phones that were nearby for a relevant period. The minimum amount of time and the minimum signal strength required to be considered a potential contact between two phones are parameters that have to be set in advance and that affect the sensitivity of the system and the number of false alarms. When someone is diagnosed with the disease, the code for the phone of that user is stored in a central database, which can be consulted regularly by the application. If a code of a phone that was in close proximity of a user phone is found in the database, during the relevant period of possible contagion, the user is warned of that fact. DP-3T was developed with a strong focus on privacy issues and is, indeed, robust against a number of different attacks [8]. In particular, the phone codes change over time, and a number of other measures were taken in order to limit risks to privacy. However, it is a fact that no system can be completely foolproof and a number of authors have argued that diverse weaknesses of the DP-3 T framework can be explored [9].

The Portuguese implementation of the DP-3T framework, the application STAY-AWAY COVID was made available in early September 2020, was widely announced and promoted by the government and, by the end of 2020, was downloaded almost three million times. If all these three million copies of the application were indeed up and running, and a significant fraction of the infections had been registered in the database, the application could have played a significant role in tracing the spread of the infection. In reality, less than 1% of known cases were registered in the system during the first four months, a factor that has severely undermined the usefulness of the application as a mechanism to stop the spread of the disease.

4 Analysis and Conclusions

At the time of this writing, we are still unable to report the final impact of the COVID-19 pandemic in Portugal, as the country is still facing an extremely high level of infections, internments and deaths. Portugal was subject to a comparatively mild first wave, in the Winter/Spring of 2020, but to a much stronger second/third wave in the Fall of 2020 and Winter 2021. In the end of January 2021, Portugal had the highest number of cases per million people of all European countries and was rapidly becoming one of the most affected countries in the world, measured by the accumulated number of deaths as a fraction of the population.

One is therefore forced to conclude that the utilization of a government-endorsed COVID-19 contact tracing application did not contribute significantly to contain the spread of the pandemic in Portugal, even though many other factors may have caused the present situation. It would be useful to investigate the reasons why the utilization of the STAYAWAY COVID application was so limited and, in particular, why so few codes have been introduced in the system, making it of limited use. Although a number of discussions about the privacy issues raised by the application took place in Portugal in the Fall of 2020, it is unlikely that those issues played a significant role in the fact that only a small number of codes were introduced. No reports were made public about the actual number of active contacts traced, nor about the effectiveness of that tracing. Worries about the number of false alarms, which could make the application cumbersome, were also raised in the Fall of 2020. However, no reports about the relevant indicators, which are anyway hard to estimate, were made available by the government. Most likely, the lack of clear protocols and procedures, coupled with the limited resources available in the national health system played a larger role in making the application irrelevant.

It is only fair to state that we do not know if the STAYAWAY COVID application could have been effective in containing the spread of COVID-19 if it had been used more extensively by the population and the national health service. Some authors have argued that smartphone-based contact tracing is an effective epidemic mitigation measure, worth being adopted by governments [1]. However, the Portuguese experience does not support these findings. There may a number of different reasons for this, including the relatively low number of application downloads, the limited usage that was made of the reporting functionality, the lack of well-defined procedures or even the low level of education of the Portuguese population. Privacy issues may also have played a role in the limited adoption of the technology but there is no evidence that these were the main roadblocks to a more widespread utilization of the official contact tracing application.

References

1. Almagor J, Picascia S (2020) Exploring the effectiveness of a COVID-19 contact tracing app using an agent-based model. Sci Rep 10(1):1–11
2. Azevedo L, Pereira MJ, Ribeiro M, Soares A (2020) Spatiotemporal modelling of COVID-19 infection risk in Portugal. PREPRINT Available at Research Square
3. Jackson C (2009) History lessons: the Asian Flu pandemic. Br J Gen Pract 59(565):622–623
4. Jeon KE, She J, Soonsawad P, Ng PC (2018) BLE beacons for internet of things applications: survey, challenges, and opportunities. IEEE Internet Things J 5(2):811–828
5. Li J, Guo X (2020) Covid-19 contact-tracing apps: a survey on the global deployment and challenges. ArXiv Preprint. arxiv:2005.03599
6. Oliveira R, Mendonça JM (2020) STAYAWAY COVID. contact tracing for COVID-19. INESC TEC Sci Soc 1(1):58–61

7. Peckham R (2020) Viral surveillance and the 1968 Hong Kong flu pandemic. J Glob Hist 15(3):444–458
8. Troncoso C, Payer M, Hubaux JP, Salathé M, Larus J, Bugnion E, Lueks W, Stadler T, Pyrgelis A, Antonioli D, Barman L, Chatel S, Paterson K, Čapkun S, Basin D, Beutel J, Jackson D, Roeschlin M, Leu P, Pereira J et al (2020) Decentralized privacy-preserving proximity tracing. ArXiv, May, 1–46
9. Vaudenay S (2020) Analysis of DP3T. IACR Cryptol Eprint Arch 2020:399

Robots That Look After Grandma? A Gerontechnology Point of View

Yeh-Liang Hsu

Abstract This chapter discusses the challenges of using robots for the care of older adults in homes. Various types of robots are introduced, including physically assisted robots, companion / social robots, and cloud-based intelligent robots. From a gerontechnology point of view, the purpose is to address how technology can be accepted as a natural part of older adults' everyday lives in their homes. Finally, we suggest that instead of designing and building one robot capable of doing many things, we should focus on building ambient intelligence to support older adults and caregivers. The "robot" itself is just a user interface for older adults and caregivers with ambient intelligence.

1 Robots That Look After Grandma

The title of this chapter is borrowed from an article published in *The Economist* in March 2019, "Robots that look after grandma." [1]. There was a subtitle. "Because of ageing, the world needs a robotics revolution. The machines don't seem ready for one…".

The article started with this statement: "Ageing and robots are more closely related than you might think." Professor Daron Acemoglu of MIT reckoned that "ageing is the biggest single influence upon how many robots a country has." The many numbers and evidence presented in the article showed that ageing creates demand for automation in two ways.

"First, to prevent output falling as more people retire, …" Aging creates demands for more industrial robots. "Second, once people have retired, they create markets for new kinds of automation, …" Aging creates demands for more service robots.

The article then went on to explain the difference between industrial robots and service robots. "For example, if an exoskeleton helps someone lift something heavy, the person still has to be there."

Y.-L. Hsu (✉)
Gerontechnology Research Center, Yuan Ze University, Taoyuan City, Taiwan
e-mail: mehsu@saturn.yzu.edu.tw

© The Author(s), under exclusive license to Springer Nature Switzerland AG 2022

M. I. A. Ferreira and M. O. Tokhi (eds.), *Towards Trustworthy Artificial Intelligent Systems*, Intelligent Systems, Control and Automation: Science and Engineering 102,
https://doi.org/10.1007/978-3-031-09823-9_9

> Industrial robots usually replace human activity, by contrast, service robots extend it.

Many researchers and companies have been developing service robots to help with the various problems caused by population aging. However, the article pointed out that, "According to the International Federation of Robotics, an estimated 20,000 robots were sold in 2018 that could realistically be described as helpful for ageing…".

Maybe the robots are not capable enough?

> Robots will have to perform a dauntingly long list of things they cannot yet do...
>
> …one robot can do only one thing, so multiple tasks would require your home to be stuffed with machines.

Price is also a problem. Currently, robot prices are relatively high, so.

> … only rich people will buy them. That may limit their wider social acceptance.

From a gerontechnology point of view, this chapter intends to extend further and discuss the challenges of using robots for caring for older adults in everyday lives.

2 Gerontechnology, the Intersection Between Technology and Aging

There are three G's, three important academic fields for the aged society. The first two G's are Gerontology and Geriatrics, which are well defined and well developed. However, the third G, Gerontechnology, can be more closely related to older adults, caregivers, and the technology industry. Facing widespread population aging, people naturally consider applying technologies to provide positive solutions in maximizing the efficiency and effectiveness of the workforce and resources for the care of older adults.

The term gerontechnology was first defined in the First International Congress on Gerontechnology, 1991, in Eindhoven, the Netherlands [2]. The International Society for Gerontechnology (ISG) was later established in1997. Gerontechnology is actually about designing for people, as defined by the ISG [3]:

"Gerontechnology: designing technology and environment for independent living and social participation of older persons in good health, comfort, and safety."

The scope of gerontechnology is a comprehensive one: health, housing, mobility, communication, leisure, work. People with different fields of expertise can find an area in gerontechnology to contribute. These are also the scope of *Gerontechnology*, the official journal of the ISG [4].

Gernotechnology is probably the only journal in the academic world that focuses on the issue, "how technology can help with the ageing world?" To find out what people do in gerontechnology, we roughly classify the 112 papers published in *Gerontechnology* from 2017 to 2020 into six categories, as shown in Table 1.

There were only seven papers published in the past four years in *Gerontechnology* about robots for senior care. The average number of downloads per paper, 170.7, is

Table 1 Papers published in *Gerontechnology* can be classified into six categories (https://journal.gerontechnology.org/)

	Category	No. papers	Avg. download per paper
1	Monitoring of vital signs/behavior patterns and platforms for health management	23	208.2
2	Technology assistance for everyday lives for older adults and caregivers	20	257.2
3	Platforms for social communication and participation of older adults	15	213.9
4	Technology interventions for enhancing physical and cognitive ability	19	216.3
5	Robots for senior care	7	170.7
6	General, non-technical issues about gerontechnolgy	28	238.2

Note The average number of downloads per paper was 224.2, all data collected 2021/02/08

also the lowest among the six categories. Observing the common terms in the titles and keywords in these seven papers to find out the main concerns of the gerontechnologists about robots, we get "acceptance," "assistive robots," "social robots," and "companion robots." So let us check them one by one.

3 Robots Used in Caring for Older Adults in Everyday Lives At Home

Physically assistive robot: What it takes for a robot to bring a glass of water?

Physically assistive robots have been developed to help older adults perform various everyday tasks, such as bringing a glass of water to the older adults. What does it take for a robot to bring a glass of water? First, the older adult has to order that water from the robot, and the robot has to understand the order. There is an interesting article published in IEEE Spectrum, and the title is "How not to order water from a robot waiter" [5]. It described a possible conversation between the human and the robot for ordering water, quoted below:

> Participant: Can I order one water?
> Robot: Yes, that is permissible.
> Participant: Okay, can you bring me one water?
> Robot: Yes, I am able to do that.
> Participant: I would like to have one water.
> Robot: Thank you for sharing that interesting fact. Please tell me your order.
> Participant: So can you do it?
> Robot: Yes. Please tell me your order.

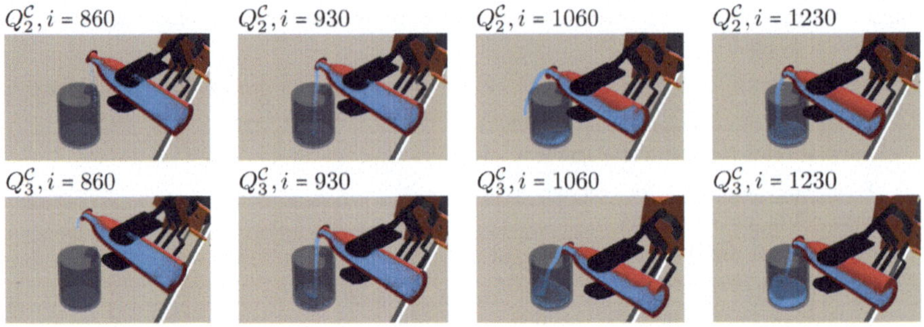

Fig. 1 "Robot motion planning for pouring liquids" [6]

Participant: My order is water—

Finally, the robot understands the command, and then the robot has to plan for a trajectory to the kitchen, locate the transparent glass and the water bottle. The next step will be a very challenging one, so there is a paper about it again. The title of the paper is "Robot motion planning for pouring liquids" [6]. As shown in Fig. 1, the robot has to hold the bottle precisely, control the angle and the water flow according to the weight of the water left in the bottle, stop pouring when the glass is full.

After that is done, the robot must hold the glass of water and navigate back to the older adult without spilling. Finally, the robot has to gently put the glass on the table close to the older adult.

Going through all these tasks, for a robot to bring a glass of water to the older adults, seems still a task belonging to that *"dauntingly long list of things they cannot yet do."*

Companion/social robot: Paro, an autonomous social robotic seal

The most famous companion/social robot for senior care should be Paro, an autonomous social robotic seal (Fig. 2) developed by the National Institute of Advanced Industrial Science and Technology (AIST), Japan [7]. Paro is the World's

Fig. 2 Paro, an autonomous social robotic seal

Most Therapeutic Robot certified by Guinness World Records in 2003, often used in non-drug intervention for people with dementia. In 2009 the US Food and Drug Administration (FDA) certified Paros as a "neurological therapeutic medical device" in biofeedback devices (class II).

The price of Paro is 6,000 USD, which is a barrier in elderly care settings. There are discussions about the cost-effectiveness of using Paro [8, 9].

> Only rich people will be able to buy it, that may limit its wider social acceptance.

At this high price, PARO must provide more value than "acting cute." There are quite a few stories, photos, and videos on the Internet showing how older adults interact with their Paros and the older adults' emotions. In these stories, Paro is often depicted as the best friend of older adults, which also creates a feeling of, loneliness.

Telepresence robot—talk to the real person behind the robot

Instead of interacting with the robotic seal, the older adults can talk to the real person behind the telepresence robot. For example, with the nickname "Dr. Robot," RP-Vita is a remote presence robot that the FDA cleared in 2013 for active patient monitoring in acute care settings [10]. The doctor can log into RP-Vita from any location with a computer and internet connection and "drive" the telepresence robot to the patient's room to interact with the patient. The screen on RP-VITA displays the image of the doctor. To make a remote doctor more effective, RP-VITA can be integrated with live patient data from the electronic medical record and connect with diagnostic devices such as otoscopes and ultrasound. This telepresence robot is already used in many hospitals.

The Giraff telepresence robot helps older adults remain in their homes and stay connected to their friends and family who are not living together. Family members, friends, doctors, and other caregivers can all log into the telepresence robot, drive it, interact with the older adults, and explore the Giraff's environment with audio and video.

There are many commercially available telepresence robots, in addition to RP-VITA and Giraff. However, the sizes (and prices) of the telepresence robots are much larger (and higher) than those of robotic vacuum machines, the only robot well-accepted by home users. If nobody logs into and drives it, the telepresence robot would only be a big machine sitting in the home.

> …one robot can do only one thing, so multiple tasks would require your home to be stuffed with machines.

For acceptance by the home users, robotic user interface (RUI) is one key issue, whether the robot should look like an animal, a human being, or a machine. A review paper published in *Gerontechnology*, "Feasibility, acceptance, and impact of socially assistive robots in non-drug interventions with people with dementia: a scoping review" collects 70 records published between 2004 and 2018 that researched feasibility, acceptance, and/or impact of socially assistive robots in non-drug interventions for people with dementia (PwD) or mild cognitive impairments (PwMCI). PwDs and PwMCIs "mostly viewed robots as non-threatening, non-obtrusive, and friendly. Positive comments centered on the robots' cute appearances or the interest in interacting with them."

PwDs and PwMCIs mostly enjoyed interacting with the robots, however, for the impact of the robots, "No consistent evidence was found for impacts on cognition, non-cognitive symptoms like agitation, anxiety, depression, wandering, apathy or other neuropsychiatric symptoms, quality of life, drug use, neuroactivity, and stress-related measurements" [11].

Cloud-based intelligent robots

With the rapid progress of natural language processing, people can directly talk to smart devices, such as talking to the mobile phone using Apple Siri or smart speakers Amazon Alexa or Google Home. Users can also dialogue with the cloud-based intelligent robot, a new type of companion robots, using natural language. There are quite a few commercially available cloud-based intelligent robots. Some are in human forms, such as Pepper developed by SoftBank, Japan (https://www.softbank.jp/en/robot/) or Zenbo (Fig. 3) by a Taiwanese company ASUS (https://zenbo.asus.com/tw/). Another approach is to add facial expressions or human movements to smart speakers to turn the smart speakers into robots. They often have embedded autonomous behaviors and can also be used in telepresence mode plus some IoT functions.

We did a usability test for this type of cloud-based intelligent robots. We used Zenbo to compare against iPhone Siri. The price of Zenbo was about 800 US dollars, which is cheaper than most iPhones. Forty-one participants aged 56–75 were recruited. They were requested to conduct the same information based tasks such as asking for the weather, asking for traffic, and function-based tasks such as turning on the light, with both Zenbo and Siri, wearing eye trackers. After that, we did a system usability scale test (SUS) with the participants. Table 2 shows the results of the SUS. Siri received significantly higher scores ($p < 0.005$) [12].

Fig. 3 Zenbo, a loud-based intelligent robot

Table 2 SUS scores for Siri and Zenbo

		No. participants	Siri	Zenbo
Sex	Male	8	69.69	62.50
	Female	33	66.77	56.52
Age	56–60	7	69.29	50.83
	61–65	17	62.83	54.85
	66–75	17	70.31	61.41
Education	College	14	68.85	53.57
	High school	24	66.05	59.07
	Primary school	3	72.50	65.83
Average scores			67.37	57.68

The participants were requested to provide feedback at the end of the SUS questionnaire. This comment came up many times: "I can do everything with Siri. Why do I need a robot?".

So back to our question, what is the best robot that looks after grandma? From a gerontechnology point of view, the whole IoT environment with sensors, actuators, and cloud servers is the most resourceful robot.

4 Building Ambient Intelligence to Support Older Adults and Caregivers

The best robot that looks after grandma can be the ambient intelligence, which means everyday environments sensitive and responsive to humans' presence. Instead of designing and building one robot capable of doing many things, we should focus on building ambient intelligence to support older adults and caregivers. This is what the other categories in gerontechnology listed in Table 1 do: Monitoring of vital signs/behavior patterns and platforms for health management; Technology assistance for everyday lives for older adults and caregivers; Technology interventions for enhancing physical and cognitive ability; Platforms for social communication and participation of the older adults. The "robot" itself is just a user interface for older adults and caregivers with ambient intelligence. After all,

> Industrial robots usually replace human activity, by contrast, service robots extend it.

The Gerontechnology Research Center (GRC) of Yuan Ze University (YZU), Taiwan, was first established in 2003. It is the pioneering research institute in the field of gerontechnology in Taiwan. As shown in Fig. 4, the core of GRC is the research and development of smart living, IoT products, with applications in gerontechnology. As a university, GRC also supports the YZU d.school, a design school in gerontechnology where we practice transdisciplinary learning. GRC supports two

Fig. 4 Gerontechnology Research Center, Yuan Ze University, Taiwan

journals. *Gerontechnology*, the official journal of the ISG, and *Gerontechnology and Service Management*, published in Chinese.

Gerontechnology research is only valuable if the research can be turned into real products for daily applications. GRC has done many industry-academia collaboration projects. In 2016, GRC moved one step forward and established a start-up company Seda GTech. In this start-up company, we take a design approach to gerontechnology, extending from creativity and prototypes in the university into products. What is very challenging for us is turning the products into sales and daily applications. With the intention of selling the products, we truly care about our target users, their needs, and the values our products can bring to them. That completes the full design cycle.

Some of the product design experience, in particular, "How the technology can be accepted as a natural part of older adults' everyday lives in the homes?" is described here.

Smart technologies that belong to the home

Figure 5 shows the IT structure of the ambient intelligence platform developed in GRC, YZU. Telehealth systems are often designed to connect patients with medical doctors and professional caregivers, who read older adults' vital sign measurements and take necessary actions. Instead, our system aims to connect older adults with children, family members, and caregivers in the home environment. Therefore we have been cautious in selecting smart technologies that belong to the homes—not necessarily high-end technologies.

Internet of things (IoT) provides an excellent tool for building ambient intelligence in the homes. However, Wi-Fi coverage is a major barrier to implementing IoT homecare products. In a survey in Taiwan in 2018, only 31.1% of older adults age 65

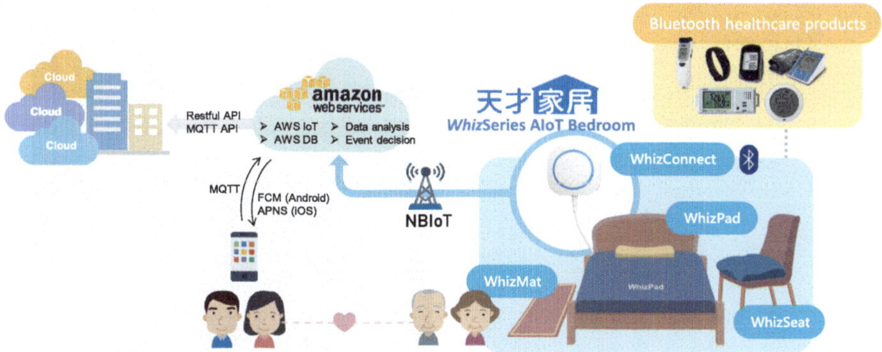

Fig. 5 IT structure of the ambient intelligence platform developed in GRC, YZU

or older used the Internet. Most of them used mobile phones to access the Internet, and only 8.6% used Wi-Fi. For the older adults who do not use Wi-Fi, paying a high monthly fee for Internet service just for IoT homecare products is not acceptable. For older adults who have Wi-Fi at home, connecting the IoT products to the home Wi-Fi AP has been a significant usability issue in our experience. Moreover, Wi-Fi coverage is often poor in bedrooms.

The center of the platform in Fig. 5 is *Whiz* Connect, the BlueTooth to narrowband IoT (NB-IoT) gateway. *Whiz* Connect uses a sim card to connect directly to mobile phone base stations and transmits data to Amazon Web Services (AWS) using the standard IoT MQTT protocol. With NB-IoT, the care platform can be used anywhere plug-and-play without setting up Wi-Fi. We believe this is the last mile to bring IoT care products to the homes.

Bluetooth and mobile devices are the two most commonly used technologies by home users. So we certainly have to utilize them. Bluetooth low energy (BLE) technology has gradually become the mainstream of home product data transmission. The two features of power-saving and broadcast messages are very conducive to homecare applications. With *Whiz*Connect, caregivers can combine commercially available BLE care products (such as wearable devices, blood pressure/glucose meter, and temperature/humidity sensors) according to their care needs, include them into the IoT ecosystem to build a customized system.

We also developed our own *Whiz*-Series smart products, as shown in Fig. 8. Instead of developing more technological devices, we chose to implement sensing, BLE communication, and machine learning capabilities into mattresses, floor mats, and chair cushions, which are already familiar to home users.

For example, *Whiz*Pad (Fig. 6) is a comfortable mattress made by temperature-sensitive foam to prevent pressure ulcers. *Whiz*Pad is capable of motion sensing with 30 sensing areas. We used machine learning to classify the sleep postures to capture the intention of leaving the bed to prevent hospital falls. The Neural Network model is deployed on the chip in the control box of *Whiz*Pad. Then our system can provide the so-call "three-stage leave bed alert." When the patient at high risk of falling sits

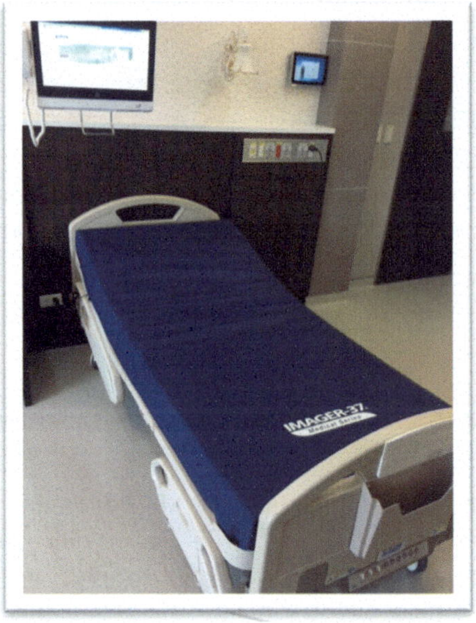

Fig. 6 *Whiz*Pad is a comfortable mattress capable of motion sensing

up on the bed, *Whiz*Pad sends the first alert to the nursing station or mobile phone carried by the nurse, second alert when the patient sits at the edge of the bed, and third alert when the patient leaves the bed. The alert function can be set up according to the condition of the individual patient. Based on the data generated by *Whiz*Pad, sleep monitoring, and long-term life pattern recognition functions are also developed.

*Whiz*Pad has been widely used in hospitals and nursing homes in Taiwan and is now entering the home market. Among all the technical functions, the feature that attracts caregivers the most is that *Whiz*Pad is also a BLE device powered by two AAA batteries and does not need to plug into the wall.

What about robots?

As mentioned earlier, from a gerontechnology point of view, the whole IoT environment with sensors, actuators, and cloud servers is the most resourceful robot. The "robot" itself is just a user interface for older adults and caregivers with ambient intelligence. On the other hand, the mobile device is the best user interface, or the only user interface acceptable for home users.

Realizing it may not be possible to design a new user interface for home users to replace mobile devices, we designed "WOBOT" in 2014 [13], which is a simple robotic frame that turns the mobile device put on it into a robot (Fig. 7). When stationary, WOBOT appears as a photo frame with two hands holding the mobile device. With three degrees of freedom in motion, WOBOT brings some of the animated characters' features into real motions.

In telepresence mode, the remote user manipulates WOBOT to show cartoon-like facial expressions and body motions while video conferencing with the local user, the

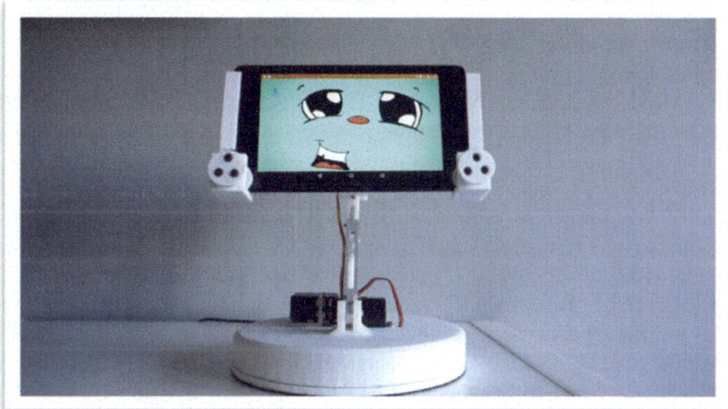

Fig. 7 WOBOT is a robotic frame that turns mobile devices into robots

older adult staying with the robot in their home environment. By providing both verbal and nonverbal interpersonal communication elements, the robot can better serve as the family members' or caregivers' avatar for expressing their care to the older adults at home. In autonomous mode, Apps on the mobile device display facial expressions, music, or video clips, while transmitting commands to the robotic platform through Bluetooth to perform requested motions. WOBOT then becomes a companion robot that offers interactive communication with older adults. Reminders, alert messages are sent to the mobile device and displayed dynamically with robotic motion. Cloud-based natural language processing and IoT functions (which were not available in 2014) can also be added for this purpose.

Years later, there are many "robotic mobile phone stands" commercially available. Prices are much lower than typical robots too.

Ambient intelligence built upon smart technologies that belong to the homes, integrated with mobile devices on a dynamic robotic frame as the user interface, can be the best form of robots that can look after grandma.

References

1. Robots that look after grandma. The Economist, March 2019
2. Bouma H, Graafmans JA (eds) (1992) Gerontechnology, vol 3. IOS Press
3. International Society for Gerontechnology [online]. https://www.gerontechnology.org/about.html. Accessed 09 Feb 2021. Robots that look after grandma? A gerontechnology point of view 13. http://www.GTechSeda.com.tw/
4. Gerontechnology. Official Journal of the International Society for Gerontechnology [online]. https://journal.gerontechnology.org/about.aspx. Accessed 09 Feb 2021
5. Ackerman E (2018) How not to order water from a robot waiter. IEEE Spectrum [online]. https://spectrum.ieee.org/automaton/robotics/artificial-intelligence/how-not-to-order-water-from-a-robot-waiter. Accessed 09 Feb 2021

6. Pan Z, Park C, Manocha D (2016) Robot motion planning for pouring liquids. In: Proceedings of the international conference on automated planning and scheduling, vol 26, no 1
7. PARO therapeutic robot [online]. http://www.parorobots.com/. Accessed 09 Feb 2021
8. Mervin MC, Moyle W, Jones C, Murfield J, Draper B, Beattie E, Thalib L (2018) The cost-effectiveness of using PARO, a therapeutic robotic seal, to reduce agitation and medication use in dementia: findings from a cluster–randomized controlled trial. J Am Med Dir Assoc 19(7):619–622
9. Hung L, Liu C, Woldum E, Au-Yeung A, Berndt A, Wallsworth C, Chaudhury H (2019) The benefits of and barriers to using a social robot PARO in care settings: a scoping review. BMC Geriatr 19(1):1–10
10. Owano N (2013) FDA gives green light to RP-VITA hospital robot, Phys.org [online]. https://phys.org/news/2013-01-fda-green-rp-vita-hospital-robot.html. Accessed 09 Feb 2021
11. Wasić C, Pendergrass A, Böhme H, Bahrmann F, Graessel E (2020) Feasibility, acceptance, and impact of socially assistive robots in non-drug interventions with people with dementia: a scoping review. Gerontechnology 20:1–25
12. Bai L, Cheng CK, Lin EJ, Chen SY, Chang IY, Hsu YL (2018) User experience test of companion robot and its innovative usage among older adults-use Zenbo as an example. J Gerontechnol Serv Manag 6(3):265–282
13. Sebastian J, Hsu YL, Lu JM (2014) Creation of a 'Caricature Robot' for social inclusion of older adults. Gerontechnology 13(2):278

Making Maximally Ethical Decisions via Cognitive Likelihood and Formal Planning

Michael Giancola, Selmer Bringsjord, Naveen Sundar Govindarajulu, and Carlos Varela

Abstract This chapter attempts to give an answer to the following question: Given an obligation and a set of potentially-inconsistent, ethically-charged beliefs, how can an artificially-intelligent agent ensure that its actions maximize the likelihood that the obligation is satisfied? Our approach to answering this question is in the intersection of several areas of research, including automated planning, reasoning with uncertainty, and argumentation. We exemplify our reasoning framework in a case study based on the famous, heroic ditching of US Airways Flight 1549, an event colloquially known as the "Miracle on the Hudson."

1 Introduction

This chapter attempts to give an answer to the following question: Given an obligation and a set of potentially-inconsistent, ethically-charged beliefs, how can an artificially-intelligent agent ensure that its actions maximize the likelihood that the obligation is satisfied? We present a framework for producing such agents. Our framework includes components from several intersecting areas of research, includ-

M. Giancola (✉)
Rensselaer AI and Reasoning (RAIR) Lab, Department of Computer Science, Department of Cognitive Science, Rensselaer Polytechnic Institute, Troy, NY, USA
e-mail: mike.j.giancola@gmail.com

S. Bringsjord
Rensselaer AI and Reasoning Lab, Department of Computer Science, Department of Cognitive Science, Rensselaer Polytechnic Institute, Troy, NY, USA
e-mail: selmer.bringsjord@gmail.com

N. S. Govindarajulu
Rensselaer AI and Reasoning Lab, Rensselaer Polytechnic Institute, Troy, NY, USA
e-mail: naveen.sundar.g@gmail.com

C. Varela
Worldwide Computing Lab, Department of Computer Science, Rensselaer Polytechnic Institute, Troy, NY, USA
e-mail: cvarela@cs.rpi.edu

© The Author(s), under exclusive license to Springer Nature Switzerland AG 2023
M. I. A. Ferreira and M. O. Tokhi (eds.), *Towards Trustworthy Artificial Intelligent Systems*, Intelligent Systems, Control and Automation: Science and Engineering 102,
https://doi.org/10.1007/978-3-031-09823-9_10

ing automated planning, reasoning with uncertainty, and argumentation. Therefore, before we proceed with a description of our framework, we begin with a discussion of technical preliminaries which will enable the construction of our framework. This discussion includes a review of previously published concepts and new content. The former includes a review of *cognitive calculi* (Sects. 2.1, 2.2) and *automated planning* (Sect. 3), while the latter includes an introduction to *cognitive likelihood* (Sect. 2.3).

We then present our framework in Sect. 4, and its application in a case study based on US Airways Flight 1549 and its "miraculous" saving, in Sect. 5. Finally, we discuss some related work (Sect. 6) and conclude (Sect. 7).

2 Cognitive Calculi

Our[1] approach to formally capturing ethics so as to install it in an artificial agent has long been grounded in the use of cognitive calculi (used e.g., in [1], the precursor to this book chapter, and [2, 3]). In short, a cognitive calculus is a multi-operator quantified intensional logic built to capture all propositional attitudes in human cognition.[2] While a longer discussion of precisely what a cognitive calculus is is out of scope, the interested reader is pointed to Appendix A in Bringsjord et al. [6].

For purposes of this chapter, it's specifically important to note that a cognitive calculus consists of *essentially* two components: (1) multi-sorted n^{th}-order logic with modal operators for modeling cognitive attitudes (e.g., knowledge **K**, belief **B**, and obligation **O**[3]) and (2) inference schemata that—in the tradition of proof-theoretic semantics—express the semantics of the modal operators. In particular, we will utilize the Inductive Deontic Cognitive Event Calculus ($\mathcal{IDCEC}$) in the work described herein. We next review a predecessor of $\mathcal{IDCEC}$, the (deductive) Deontic Cognitive Event Calculus ($\mathcal{DCEC}$).

2.1 Deontic Cognitive Event Calculus

$\mathcal{DCEC}$ is fully captured in the following two boxes, titled $\mathcal{DCEC}$ **Signature** and $\mathcal{DCEC}$ **Inference Schemata**. They contain the sorts, function signatures, grammar, and inference schemata which comprise $\mathcal{DCEC}$. Notice that, while cognitive calculi can be constructed from n^{th}-order logic (for any value of $n \geq 0$), the standard[4] $\mathcal{DCEC}$ is built with a core of first-order logic.

[1] This collective refers to the RAIR Lab, of which the first three authors are members.

[2] For information about such attitudes, see [4]; for a wonderful catalogue of all the major categories of human cognition, from *perceiving* to *fearing* to *remembering* to *saying* and beyond, see [5].

[3] See the box titled $\mathcal{DCEC}$ **Signature** within Sect. 2.1 for the rest of the modal operator descriptors.

[4] Note that several variants of $\mathcal{DCEC}$ have been formalized and deployed. For example, [2] uses $\mathcal{DCEC}^*$, a version of $\mathcal{DCEC}$ which allows for the formal modeling of self-reflective agents.

Also, an automated reasoner for $\mathcal{DCEC}$–ShadowProver [7]—has been created, is available, and is under active development.

$\mathcal{DCEC}$ Signature

$$S ::= \textsf{Agent} \mid \textsf{ActionType} \mid \textsf{Action} \sqsubseteq \textsf{Event} \mid \textsf{Moment} \mid \textsf{Fluent}$$

$$f ::= \begin{cases} action : \textsf{Agent} \times \textsf{ActionType} \rightarrow \textsf{Action} \\ initially : \textsf{Fluent} \rightarrow \textsf{Formula} \\ holds : \textsf{Fluent} \times \textsf{Moment} \rightarrow \textsf{Formula} \\ happens : \textsf{Event} \times \textsf{Moment} \rightarrow \textsf{Formula} \\ clipped : \textsf{Moment} \times \textsf{Fluent} \times \textsf{Moment} \rightarrow \textsf{Formula} \\ initiates : \textsf{Event} \times \textsf{Fluent} \times \textsf{Moment} \rightarrow \textsf{Formula} \\ terminates : \textsf{Event} \times \textsf{Fluent} \times \textsf{Moment} \rightarrow \textsf{Formula} \\ prior : \textsf{Moment} \times \textsf{Moment} \rightarrow \textsf{Formula} \end{cases}$$

$$t ::= x : S \mid c : S \mid f(t_1, \ldots, t_n)$$

$$\phi ::= \begin{cases} q : \textsf{Formula} \mid \neg\phi \mid \phi \wedge \psi \mid \phi \vee \psi \mid \forall x : \phi(x) \mid \exists x : \phi(x) \\ \mathbf{P}(a, t, \phi) \mid \mathbf{K}(a, t, \phi) \mid \mathbf{S}(a, b, t, \phi) \mid \mathbf{S}(a, t, \phi) \\ \mathbf{C}(t, \phi) \mid \mathbf{B}(a, t, \phi) \mid \mathbf{D}(a, t, \phi) \mid \mathbf{I}(a, t, \phi) \\ \mathbf{O}(a, t, \phi, (\neg) happens(action(a^*, \alpha), t')) \end{cases}$$

Modal Operator Descriptors:
Perceives, **K**nows, **S**ays,
Common-knowledge
Believes, **D**esires, **I**ntends, **O**ught-to

$\mathcal{DCEC}$ Inference Schemata

$$\frac{\mathbf{K}(a, t_1, \Gamma), \quad \Gamma \vdash \phi, \quad t_1 \leq t_2}{\mathbf{K}(a, t_2, \phi)} [I_{\mathbf{K}}] \qquad \frac{\mathbf{B}(a, t_1, \Gamma), \quad \Gamma \vdash \phi, \quad t_1 \leq t_2}{\mathbf{B}(a, t_2, \phi)} [I_{\mathbf{B}}]$$

$$\frac{}{\mathbf{C}(t, \mathbf{P}(a, t, \phi) \rightarrow \mathbf{K}(a, t, \phi))} [I_1] \qquad \frac{}{\mathbf{C}(t, \mathbf{K}(a, t, \phi) \rightarrow \mathbf{B}(a, t, \phi))} [I_2]$$

$$\frac{\mathbf{C}(t, \phi)\ t \leq t_1 \ldots t \leq t_n}{\mathbf{K}(a_1, t_1, \ldots \mathbf{K}(a_n, t_n, \phi) \ldots)} [I_3] \qquad \frac{\mathbf{K}(a, t, \phi)}{\phi} [I_4]$$

$$\frac{}{\mathbf{C}(t, \mathbf{K}(a, t_1, \phi_1 \rightarrow \phi_2)) \rightarrow \mathbf{K}(a, t_2, \phi_1) \rightarrow \mathbf{K}(a, t_3, \phi_2)} [I_5]$$

$$\frac{}{\mathbf{C}(t, \mathbf{B}(a, t_1, \phi_1 \rightarrow \phi_2)) \rightarrow \mathbf{B}(a, t_2, \phi_1) \rightarrow \mathbf{B}(a, t_3, \phi_2)} [I_6]$$

$$\frac{}{\mathbf{C}(t, \mathbf{C}(t_1, \phi_1 \rightarrow \phi_2)) \rightarrow \mathbf{C}(t_2, \phi_1) \rightarrow \mathbf{C}(t_3, \phi_2)} [I_7]$$

$$\frac{}{\mathbf{C}(t, \forall x.\, \phi \rightarrow \phi[x \mapsto t])} [I_8] \qquad \frac{}{\mathbf{C}(t, \phi_1 \leftrightarrow \phi_2 \rightarrow \neg\phi_2 \rightarrow \neg\phi_1)} [I_9]$$

$$\frac{}{\mathbf{C}(t, [\phi_1 \wedge \ldots \wedge \phi_n \rightarrow \phi] \rightarrow [\phi_1 \rightarrow \ldots \rightarrow \phi_n \rightarrow \phi])} [I_{10}]$$

$$\frac{\mathbf{B}(a, t, \phi)\ \ \mathbf{B}(a, t, \phi \rightarrow \psi)}{\mathbf{B}(a, t, \psi)} [I_{11a}] \qquad \frac{\mathbf{B}(a, t, \phi)\ \ \mathbf{B}(a, t, \psi)}{\mathbf{B}(a, t, \phi \wedge \psi)} [I_{11b}]$$

$$\frac{\mathbf{S}(s, h, t, \phi)}{\mathbf{B}(h, t, \mathbf{B}(s, t, \phi))} [I_{12}] \qquad \frac{\mathbf{I}(a, t, happens(action(a^*, \alpha), t'))}{\mathbf{P}(a, t, happens(action(a^*, \alpha), t'))} [I_{13}]$$

$$\frac{\mathbf{B}(a, t, \phi)\ \ \mathbf{B}(a, t, \mathbf{O}(a, t, \phi, \chi))\ \ \mathbf{O}(a, t, \phi, \chi)}{\mathbf{K}(a, t, \mathbf{I}(a, t, \chi))} [I_{14}]$$

2.2 Inductive Deontic Cognitive Event Calculus

$\mathcal{DCEC}$ employs no uncertainty system (e.g., probability measures, *strength factors*, or likelihood measures) and hence is purely deductive. Therefore, as we wish to enable our agents to reason about situations involving uncertainty, we must ultimately utilize the *Inductive* $\mathcal{DCEC}$: $\mathcal{IDCEC}$.

In general, to go from a deductive to an inductive cognitive calculus, we require two components: (1) an uncertainty system, and (2) inference schemata that delineate the methods by which inferences linking formulae and other information can be used to build formally valid arguments.

The particular uncertainty system we use herein is discussed in Sect. 2.3. The inference schemata of $\mathcal{IDCEC}$ consist of the union of the set presented in Sect. 2.1 with that in the box titled **Additional Inference Schemata for** $\mathcal{IDCEC}$. Likewise, the signature of $\mathcal{IDCEC}$ subsumes that of the deductive $\mathcal{DCEC}$; the syntax of $\mathcal{IDCEC}$ also includes the forms given in the box titled **Additional Syntax for** $\mathcal{IDCEC}$.

Additional Syntax for $\mathcal{IDCEC}$

$$\phi ::= \{ \ \mathbf{B}^{\sigma}(a, t, \phi)$$
$$\text{where } \sigma \in [-5, -4, \ldots, 4, 5]$$

Additional Inference Schemata for $\mathcal{IDCEC}$

$$\frac{\mathbf{P}(a, t_1, \phi), \quad \Gamma \vdash t_1 < t_2}{\mathbf{B}^4(a, t_2, \phi)} \; [I^s_{\mathbf{P}}]$$

$$\frac{\mathbf{B}^{\sigma_1}(a, t_1, \phi_1), \ldots, \mathbf{B}^{\sigma_m}(a, t_m, \phi_m), \{\phi_1, \ldots, \phi_m\} \vdash \phi, \{\phi_1, \ldots, \phi_m\} \nvdash \zeta, \Gamma \vdash t_i < t}{\mathbf{B}^{min(\sigma_1, \ldots, \sigma_m)}(a, t, \phi)} \; [I^s_{\mathbf{B}}]$$
$$\text{where } \sigma \in [0, 1, \ldots, 4, 5]$$

$$\frac{}{\mathbf{C}(t, \mathbf{B}^{-\sigma}(a, t, \phi) \leftrightarrow \mathbf{B}^{\sigma}(a, t, \neg\phi))} \; [I^s_{\neg}]$$

Briefly, $\mathbf{B}^{\sigma}(a, t, \phi)$ denotes that agent a at time t believes ϕ with uncertainty σ. We justify in the next section the range of values for σ.

The first inference schema allows agents to infer evident beliefs ($\sigma = 4$, as defined in the next section) from what they perceive.[5] The second schema allows agents to infer a belief that is provable from the beliefs they currently assert, so long as the belief set is not inconsistent. In practice, we usually check that the belief set is consistent by attempting to prove a reserved propositional atom ζ which does not appear anywhere else; hence, ζ can only be proved if $\{\phi_1, \ldots, \phi_m\}$ is inconsistent.[6]

[5] That is, what they perceive *externally*. We allow agents to infer *certain* beliefs on the basis of *internal* perceptions, but need not delve into this further for the purposes of the present work. $\mathcal{IDCEC}$ used herein makes no provision for these two modes of perception.

Table 1 The 11 Cognitive Likelihood values

Numerical	Linguistic
5	CERTAIN
4	EVIDENT
3	OVERWHELMINGLY LIKELY = BEYOND REASONABLE DOUBT
2	LIKELY
1	MORE LIKELY THAN NOT
0	COUNTERBALANCED
−1	MORE UNLIKELY THAN NOT
−2	UNLIKELY
−3	OVERWHELMINGLY UNLIKELY = BEYOND REASONABLE BELIEF
−4	EVIDENTLY NOT
−5	CERTAINLY NOT

The third schema specifies how uncertainty for $\neg\phi$ is derived from uncertainty for ϕ. The common knowledge construct in the third schema allows individual agents, in addition to the system, to handle both positive and negative uncertainty values.

As with $\mathcal{DCEC}$, an automated reasoner for $\mathcal{IDCEC}$–ShadowAdjudicator [1]—is under active development.

As mentioned at the opening of this subsection, in addition to inference schemata, we also require an uncertainty system. The specific uncertainty concept we employ herein is *Cognitive Likelihood*, which we now discuss.

2.3 *Cognitive Likelihood*

Our approach to quantifying the uncertainty of beliefs within cognitive calculi eschews traditional probability values in favor of *likelihood* values. The 11 likelihood values employed in this chapter are shown in Table 1.

Likelihood values can be obtained in either of two ways; both ways immediately reveal that we take likelihood to be *subjective*. The first way is to take as primitive a cognitive binary relation on formulae from the perspective of a rational agent (e.g., ϕ is *more reasonable than* ψ), and then build up formally to the partial or total order in question. This approach is first formalized in [8] and is deployed in e.g., the precursor to this chapter, [1]. Another approach, the one taken here, is to independently justify

[6] In connection with standard practice in mathematical logic, ζ functions essentially like $\perp$ or '0=1.'

each likelihood value by appeal to rational human-level cognition. We do so (briefly) next.

First, note that, because of schema $[I^s_\neg]$ presented in Sect. 2.2, we only need to define the non-negative likelihood values (as a negative likelihood value for belief in some formula ϕ is equivalent to a positive likelihood value for belief in $\neg\phi$).

That which is CERTAIN applies to propositions that a perfectly rational human-level cognizer would affirm as such—that 2+2=4 (Base-10), that $0 \neq 1$, and so on for any theorem that has been certifiably deduced from what is itself CERTAIN. Propositions that are CERTAIN needn't be mathematical in nature, only absolutely indubitable; e.g., that if something has both the properties R_1 and R_2, then it has the property R_1 qualifies.

Propositions are EVIDENT typically when they are given by immediate perception in the absence of conditions known to frequently cause illusory perception. For example, currently the lead author perceives his laptop's screen in front of him, and hence that there is such a screen in front of him is EVIDENT.

Next, as to the concept of BEYOND REASONABLE DOUBT, it has a long-standing history in many legal systems (such as e.g., the one long operative in the U.S.), being the level of argument that a prosecutor must provide in order for a court to convict the defendant. In this context, the following is required (emphasis ours):

> To establish the standard of proof *beyond reasonable doubt*, there must be a plausible explanation of the evidence that includes all of the elements of the crime and, *in addition*, there must be **no plausible explanation that is consistent with innocence**. [9] (Sect. 3.2.2. para. 12)

We next move to the center of the likelihood continuum, COUNTERBALANCED, which indicates no belief for or against a formula. From there, MORE LIKELY THAN NOT indicates the lowest level of belief above COUNTERBALANCED. Only a weak argument is required to reach this level.

Finally, we can define LIKELY as any belief whose likelihood is less than BEYOND REASONABLE DOUBT and more than MORE LIKELY THAN NOT.

3 Highly-Expressive Automated Planning

The final necessary component of our framework is an automated planner, in particular one that is fully compatible with our formalisms and their emphasis on declarative content and automated reasoning over that content in uncertain situations. The first modern automated planner was the Stanford Research Institute Problem Solver (STRIPS) [10], which produced a framework for planning upon which many modern planners are built.

The setup of a STRIPS problem is as follows. There is a set of formulae describing the initial state of the *world*, a set of *actions* which describe methods by which the planner can change the world state, and a *goal* set which denotes those formulae that the agent in question wishes to hold. The actions consist of three components: (1) a set of preconditions (formulae which must hold in order to perform the action), (2) a set of additions (formulae that will be added to the world by taking the action), and (3) a set of deletions (formulae to be removed from the world by taking the action).

The expressivity of formulae used to represent the world, actions, and goal was limited to propositional statements. For example, the goal that "the book is not on the table" could be represented by ¬On(book, table). In this work, we will need to be able to use quantified formulae, e.g., ¬∃ x On(x, table), to describe the world and goal.

The Planning Domain Definition Language (PDDL) [11] is a STRIPS-style planning language, which also supports quantification over zero-order formulae. While some quantification is supported, PDDL has serious restrictions on the syntax of formulae that can be supported. Arbitrary first-order-logic formulae are not allowed. Further, PDDL does not support modal operators such as those for *belief*, *knowledge*, or *obligation*; these are are necessary for modeling states of minds of agents. (E.g., in our case study below, we would ultimately want AI that is able to bring about "mental" goals, such as a pilot's believing that such and such a course of action is feasible.) Reasoning with such mental states is crucial in ethically charged situations.[7] Formulae of the following nature, which require the ability to nest such modal operators, cannot be expressed in PDDL:

Alice *believes* that all pilots *believe*, before entering a cockpit, that they *know* ϕ.

$$\mathbf{B}(alice, t, \forall x\ \exists t_0 t_1\ \mathbf{B}(x, t_0, \mathbf{K}(x, t_0, \phi)) \wedge EntersCockpit(x, t_1) \wedge t_0 < t_1)$$

Another major limitation of the PDDL family of languages is that they require a finite and fixed universe of objects to be specified beforehand. In many uncertain situations, this is not realistic, as the number of relevant objects and entities will be unknown. Consider a situation in which a firefighting robot has to enter a building with the goal of rescuing any humans in the building. The agent has no prior knowledge of the number of humans in the building. PDDL languages are not directly amenable to modeling such situations.

Overall, then, we need a planning formalism (with an associated automated planner) that can handle arbitrary formulae for describing the world, states of minds, and an unknown set of objects. For a planner with such capabilities, we turn to Spectra [13], a STRIPS-style planner which can be integrated with reasoners for cognitive calculi [7]. While there is an efficiency disadvantage in using a more expressive planning formalism, efficiency gains in reasoning with cognitive calculi can be transferred to efficiency gains in Spectra.

4 Selecting Plans Using Cognitive Likelihood

In our framework, agents are given the following: (1) an obligation, (2) knowledge regarding the conditions required to satisfy the obligation, and (3) a set of (potentially

[7] See [12] for an example of an ethically-charged situation in which the ascription of mental states is crucial to the success of the AI agents used.

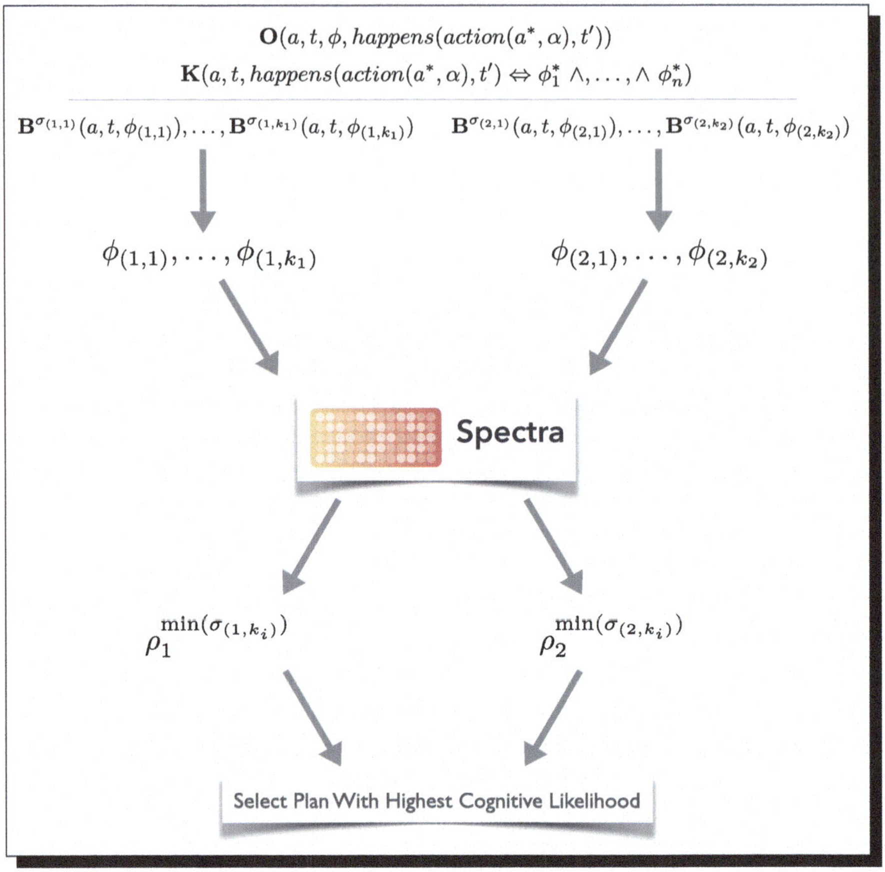

Fig. 1 A framework for selecting maximally ethical plans. This diagram shows two belief subsets, from which Spectra generates one plan each. More generally, there could be an arbitrary number of belief subsets, as well as an arbitrary number of plans generated. However, this is not shown in order to simplify the illustration

inconsistent) ethically-charged beliefs regarding actions the agent can take to affect the status of the obligation. The agents make maximally ethical decisions by taking a course of action which maximizes the agent's belief that the obligations will stay (or become) satisfied.

The decision-making framework is outlined pictorially in Fig. 1. An agent a is obligated to perform some action α, given that it believes some precondition ϕ holds. It also knows the conditions that will enable α to happen (in Fig. 1, $\phi_1^*, \ldots, \phi_n^*$).

Next, a has a set of beliefs regarding formulae pertinent to its obligations. Various subsets of those beliefs (in Fig. 1, $\phi_{(1,1)}, \ldots, \phi_{(1,k_1)}$ and $\phi_{(2,1)}, \ldots, \phi_{(2,k_2)}$), are passed to Spectra, with the goal of generating plans which cause α to occur. We assign each of those plans a likelihood based on the likelihood of the weakest belief required to generate the plan. Finally, we select the plan with the highest likelihood as the one to enact.

5 Case Study: The Miracle on the Hudson

To display our framework "in action," we consider two potential arguments concerning what decision should be made in the case of US Airways Flight 1549, colloquially known as the "Miracle on the Hudson." Namely, after losing thrust in both engines, the pilots had to quickly make the decision where to attempt an emergency landing, ultimately considering the following options: (a) attempt to return to LaGuardia Airport (LGA), (b) attempt to reach Teterboro Airport (TEB), or (c) attempt to ditch in the Hudson River.

5.1 Recounting US Airways Flight 1549

Our case study is based on the real-world aviation emergency colloquially known as the "Miracle on the Hudson." On January 15, 2009, US Airways Flight 1549 departed LaGuardia Airport (LGA) in New York City headed for Charlotte, North Carolina. Shortly after takeoff, while attempting to climb to cruising altitude, the plane flew into a large flock of Canada geese; this compromised both engines. In fact, both engines lost thrust, and despite multiple attempts the pilots were unable to regain thrust in either engine. Therefore, it quickly became evident to Captain "Sully" Sullenberger that an emergency landing was necessary, and in particular, that they "may end up in the Hudson [River]."[8] An air-traffic controller who was in communication with Captain Sullenberger gave him landing options at LaGuardia and nearby Teterboro Airport (TEB), but by the time these options were considered, neither was reachable due to the aircraft's altitude and lack of thrust in both engines. Sullenberger deftly made the executive decision to ditch in the Hudson River, thereby saving the lives of everyone onboard. Simulations of the accident have come to the conclusion that Sullenberger's decision was optimal given the preconditions [15].

5.2 The Setup

The setup of the framework for our case study is as follows. We have three agents: a_1 and a_2 will each present two inconsistent arguments regarding where the plane

[8] This quote, recorded by the in-flight cockpit voice recorder (CVR), was retrieved from the NTSB Accident Report [14].

should be landed (in the following subsection), and a^* is the adjudicator who will decide which argument and plan to proceed with.

We denote the moment after the plane flew into the flock of geese as t^*. At that time, a^* believes there is an emergency, and consequently, the agent is obligated to ensure that the landing site it selects is safe.

$$\mathbf{B}(a^*, t^*, emergency)$$
$$\mathbf{O}(a, t^*, emergency, happens(action(a^*, ensure\_safe(landing\_site)), t^{*'}))$$

The agent also knows the conditions required for a landing site to be safe: it must be close enough to reach, long and wide enough, and far enough from people, as without thrust, the pilots' ability to maneuver the plane will be more limited than usual.

$$\mathbf{K}\Big(a, t^*, happens\Big(action(a^*, ensure\_safe(landing\_site)), t^{*'}\Big) \Leftrightarrow \vdash Safe(landing\_site)\Big)$$

$$\mathbf{K}\left(a, t^*, \forall \ell\ Safe(\ell) \Leftrightarrow \bigwedge \left\{ \begin{array}{c} CloseEnough(\ell), \\ LongEnough(\ell), \\ WideEnough(\ell), \\ FarEnoughFromPeople(\ell) \end{array} \right\}\right)$$

5.3 The Arguments

We next give two arguments in favor of selecting different landing locations based on the conditions of Flight 1549, then show how our framework would generate plans for each, and finally, how it would select a plan to execute.

5.3.1 Argument 1

The first agent argues for the following two statements:

$$\mathbf{B}^3(a_1, t^*, CloseEnough(lga))$$
$$\mathbf{B}^1(a_1, t^*, FarEnoughFromPeople(lga))$$

The first formula states that it is OVERWHELMINGLY LIKELY that LaGuardia Airport was close enough for the pilots to successfully land there. This is justified by the several studies and simulations performed since the event which identified many feasible trajectories to enable landing at several different runways at LGA (e.g., see [15, 16]).

The second states that it is MORE LIKELY THAN NOT that LGA is far enough from people to ensure a safe landing despite the conditions (i.e., loss of thrust in both engines at low altitude, occurring in—to quote Captain Sullenberger—"a highly developed, metropolitan area" [17]). The likelihood is necessarily weak, as the corresponding justification is weak. As there is no data to go on, one can only speculate that based on the Captain's training, and Air Traffic Control's ability to clear a runway in time, that it is possible that the plane could have been landed at LGA without harming anyone on the ground.

5.3.2 Argument 2

The second argument asserts the following two statements:

$$\mathbf{B}^{-2}(a_2, t^*, \mathit{CloseEnough}(\mathit{teb}))$$

$$\mathbf{B}^{-2}(a_2, t^*, \mathit{FarEnoughFromPeople}(\mathit{lga}))$$

The first statement, that it is UNLIKELY that Teterboro Airport is close enough, was asserted without justification by Captain Sullenberger in the public hearing on the accident [17]. He likely intended to imply an implicit justification that it was obvious to him based on his experience as a pilot.

Second, a_2 asserts that it is UNLIKELY that LGA was far enough from people to ensure a safe landing. Note that this belief is directly inconsistent with a belief of a_1; namely, $\mathbf{B}^1(a_1, t^*, \mathit{FarEnoughFromPeople}(\mathit{lga}))$. Again, this is justified by a statement provided by Captain Sullenberger during the public hearing:

> Looking at where we were and how much time, altitude, and distance would be required to turn back toward LaGuardia and then fly toward LaGuardia, I determined quickly that that was going to be problematic, and it would not be a realistic choice, and I couldn't afford to be wrong. [17]

It is clear that, had Captain Sullenberger chosen to attempt a landing at LGA, he would've risked the lives of people at and around LGA, *in addition* to the inevitable risk already imposed on those in the plane by the emergency.

5.4 *The Framework, Applied*

We now present the application of our framework, in order to adjudicate these clearly inconsistent arguments[9] and generate a plan. First, the content of each agent's beliefs are passed separately to Spectra. Therefore, the first agent passes:

[9] We note that when we refer to *arguments* being inconsistent, this is to say that the arguments each assert a set of formulae, and from the union of those sets, a contradiction can be deduced.

$$\mathit{CloseEnough}(lga)$$
$$\mathit{FarEnoughFromPeople}(lga)$$

and the second agent passes:

$$\neg \mathit{CloseEnough}(teb)$$
$$\neg \mathit{FarEnoughFromPeople}(lga)$$

In order to generate plans, Spectra is given the following actions:

```
(define-action considerRunwayLanding [?r]
  {:preconditions [(CloseEnough ?r)
                   (FarEnoughFromPeople ?r)
                  ]

   :additions     [(LongEnough ?r)
                   (WideEnough ?r)
                  ]

   :deletions     [ ]
  }
)


(define-action considerTerrainLanding []
  {:preconditions [(not (and (Safe lga) (Safe teb)))
                  ]

   :additions     [(CloseEnough hud)
                   (LongEnough  hud)
                   (WideEnough  hud)
                   (FarFromPeopleEnough hud)
                  ]

   :deletions     [ ]
  }
)
```

The first action requires that, in order to consider landing at a particular runway, the ethically-charged propositions are first satisfied. It then adds that the runway satisfies basic requirements. The idea here is that, if implemented "for real," Spectra would be integrated with systems which could provide the necessary data, i.e., the length and width of the runway being considered, and the length and width required.

The second action allows the planner to consider off-runway landing options *only* if the runway options have been exhausted (that is, it has been determined that none of them meet the imposed safety requirements). As with the first action, Spectra would need to be integrated with another system. In this case, our simulation assumes Spectra would have access to a vision-based landing-site detection system, such as that presented in [18]. Shen et al. specify that (emphasis ours):

> A landing-site is considered safe only if its surface is *smooth* and if its *length* and *width* are adequate. [18] (pg. 295)

At the public hearing, Sullenberger stated that (emphasis ours):

> [Other than LGA or TEB,] the only place in a highly developed, metropolitan area, *long enough*, *wide enough*, *smooth enough* to land was the river. [17] (pg. 25)

Hence we can confidently say that, were Shen et al.'s system integrated with Spectra in this case, the river would have been the only landing option returned.

Each agent's input to Spectra returns a single plan. The former indicates that the pilot can land LGA, as all safety requirements have been satisfied. Alternatively, the latter is able to prove that neither LGA nor TEB are safe options, and hence seeks out off-runway options, and finds the Hudson as the only option.

Finally, note that the weakest likelihood used by agent 1 is MORE LIKELY THAN NOT (= 1) and the weakest of agent 2's argument is LIKELY (= 2). Hence the framework would conclude that agent 2's argument, and corresponding plan, are to be used.

6 Related Work

The most directly related work is [1], a precursor to this book chapter, in which the authors first used uncertainty-infused cognitive calculi to reason about the ethical decision-making involved in the Miracle on the Hudson. Like this paper, it employed an uncertainty system: it used strength factors [8] whereas the present work uses cognitive likelihood. Also, one shortcoming of the prior paper was that the AI agent featured there was given the Hudson as a landing option from the outset. In our subsequent work, reported herein, we show how an AI can creatively[10] find the Hudson on its own (i.e., by deploying a vision-based landing-site detection system such as that presented in [18]).

---

[10] A precise account of what sort of creativity is used by the AI in reasoning to the Hudson as a landing area is beyond the present paper. A number of increasingly impressive types of creativity in an artificial agent are laid out in [19]. Note that at least one expert on AI and creativity, Cope, would classify the AI described in the present paper as creative; see [20].

Giancola et al. [1] (and consequently, this work as well) was inspired by [21]. This paper presented a framework intended to ensure that autonomous systems[11] make certifiably ethically correct decisions. In particular, when no completely ethical decision is available (i.e., each possible decision will violate at least one ethical principle), they formally verified that their system will always pick the "least unethical" choice available. They achieve this verification using exhaustive model checking over the configuration of the world state as well as the ethical considerations in play.

However, the work has several shortcomings by our metrics; these deficiencies are expounded and overcome in [1]. Briefly, (1) the model-checking process used in [21] is too slow for practical applications, (2) the formalism used, a Belief-Desire-Intention (BDI) language, is too inexpressive, and (3) there is no conception of uncertainty in their system. Giancola et al. [1] also elaborates on the infeasibility of formal verification based on model checking in general.

Another area of prior work is the creation of linear models for evaluating potential landing sites [16] and trajectories [15] for Flight 1549. We applaud these works, as they produced systems which could have given the pilots actionable data (i.e., trajectories and landing sites to use to avoid ditching in the Hudson). We envision that ultimately, a system could be engineered which integrates the work herein with the linear models of [15, 16], in order to generate plans that reflect relevant data gathered by an aircraft's instruments as well as the pilots' beliefs and ethical concerns.

7 Conclusion

We presented a framework for AI agents to formally produce a maximally ethical plan based on uncertain, ethically-charged beliefs. We then showed how this framework could potentially be deployed using US Airways Flight 1549 as a case study. As with much logic-based AI research, one major area of future work is developing methods of integrating the framework proposed herein with the necessary components to enable practical usage (e.g., vision systems, aircraft sensors, and connectionist AI systems such as neural networks).

Acknowledgements The authors are grateful to AFOSR for its longstanding support of Bringsjord et al.'s development of cutting-edge formalisms and automated reasoning and planning systems for high levels of computational intelligence. We deeply thank ONR, both for their support of our current work on belief adjudication (as part of our development of AI that surmounts Arrow's Impossibility Theorem and related theorems), and for their past support of our R&D in robot/machine/AI ethics (primarily under a MURI to advance the science and engineering of moral competence in robots).

[11] We use the term 'system' in reference to the work of Dennis et al. as this is the term they use in their own work. However, in keeping with the terminology of standard textbooks in AI [22, 23], we use the term 'agent' in our work.

References

1. Giancola M, Bringsjord S, Govindarajulu NS, Varela C (2020) Ethical reasoning for autonomous agents under uncertainty. In: Tokhi M, Ferreira M, Govindarajulu N, Silva M, Kadar E, Wang J, Kaur A (eds) Smart living and quality health with robots, proceedings of ICRES 2020, CLAWAR, London, UK, pp 26–41. http://kryten.mm.rpi.edu/MG_SB_NSG_CV_LogicizationMiracleOnHudson.pdf. The Shadow Adjudicator system can be obtained here: https://github.com/RAIRLab/ShadowAdjudicator
2. Bringsjord S, Govindarajulu N, Thero D, Si M (2014) Akratic robots and the computational logic thereof. In: Proceedings of ETHICS 2014 (2014 IEEE symposium on ethics in engineering, science, and technology), Chicago, IL, pp 22–29. http://ieeexplore.ieee.org/xpl/mostRecentIssue.jsp?punumber=6883275, IEEE Catalog Number: CFP14ETI-POD. (Papers from the Proceedings can be downloaded from IEEE at URL provided here)
3. Govindarajulu N, Bringsjord S (2017a) On automating the doctrine of double effect. In: Sierra C (ed) Proceedings of the twenty-sixth international joint conference on artificial intelligence (IJCAI-17), international joint conferences on artificial intelligence, pp 4722–4730. https://doi.org/10.24963/ijcai.2017/658
4. Nelson M (2015) Propositional attitude reports. In: Zalta E (ed) The Stanford Encyclopedia of philosophy. https://plato.stanford.edu/entries/prop-attitude-reports
5. Ashcraft M, Radvansky G (2013) Cognition, 6th edn. Pearson, London, UK
6. Bringsjord S, Govindarajulu NS, Licato J, Giancola M (2020) Learning *Ex Nihilo*. In: Danoy G, Pang J, Sutcliffe G (eds) GCAI 2020, 6th global conference on artificial intelligence (GCAI 2020), EasyChair, EPiC series in computing, vol 72, pp 1–27. https://doi.org/10.29007/ggcf; https://easychair.org/publications/paper/NzWG
7. Govindarajulu N, Bringsjord S, Peveler M (2019) On quantified modal theorem proving for modeling ethics. In: Suda M, Winkler S (eds) Proceedings of the second international workshop on automated reasoning: challenges, applications, directions, exemplary achievements (ARCADE 2019). Electronic proceedings in theoretical computer science, vol 311. Open Publishing Association, Waterloo, Australia, pp 43–49. http://eptcs.web.cse.unsw.edu.au/paper.cgi?ARCADE2019.7.pdf; The Shadow Prover system can be obtained here https://naveensundarg.github.io/prover/
8. Govindarajulu NS, Bringsjord S (2017b) Strength factors: an uncertainty system for quantified modal logic. In: Belle V, Cussens J, Finger M, Godo L, Prade H, Qi G (eds) Proceedings of the IJCAI workshop on "Logical Foundations for Uncertainty and Machine Learning" (LFU-2017). Melbourne, Australia, pp 34–40. http://homepages.inf.ed.ac.uk/vbelle/workshops/lfu17/proc.pdf
9. Ho HL (2015) The legal concept of evidence. In: Zalta EN (ed) The Stanford Encyclopedia of philosophy, winter, 2015th edn. Stanford University, Metaphysics Research Lab, Stanford
10. Fikes RE, Nilsson NJ (1971) STRIPS: a new approach to the application of theorem proving to problem solving. Artif Intell 2(3–4):189–208
11. Mcdermott D, Ghallab M, Howe A, Knoblock C, Ram A, Veloso M, Weld D, Wilkins D (1998) PDDL—the planning domain definition language. Tech. Rep. CVC TR-98-003, Yale Center for Computational Vision and Control
12. Bringsjord S, Govindarajulu N, Giancola M (2021) Automated argument adjudication to solve ethical problems in multi-agent environments. Paladyn J Behav Robotics. 12(1):310–335. https://doi.org/10.1515/pjbr-2021-0009
13. Naveen Sundar G (2017) Spectra. https://naveensundarg.github.io/Spectra/
14. Hersman DA, Hart CA, Sumwalt RL (2010) Loss of thrust in both engines after encountering a flock of birds and subsequent ditching on the Hudson River. Accident Report NTSB/AAR-10/03, National Transportation Safety Board (NTSB)
15. Paul S, Hole F, Zytek A, Varela CA (2017) Flight trajectory planning for fixed wing aircraft in loss of thrust emergencies. In: dynamic data-driven application systems (DDDAS 2017). Cambridge, MA. http://arxiv.org/abs/1711.00716

16. Atkins E (2010) Emergency landing automation aids: an evaluation inspired by US Airways Flight 1549. In: AIAA Infotech@Aerospace 2010. https://doi.org/10.2514/6.2010-3381
17. National Transportation Safety Board (NTSB) (2009) Transcript—public hearing day 1. Landing of US Airways Flight 1549, Airbus A320, N106US, in the Hudson River. https://data.ntsb.gov/Docket?NTSBNumber=DCA09MA026
18. Shen YF, Rahman ZU, Krusienski D, Li J (2013) A vision-based automatic safe landing-site detection system. IEEE Trans Aerosp Electron Syst 49(1):294–311
19. Bringsjord S, Sen A (2016) On creative self-driving cars: hire the computational logicians, fast. Appl Artif Intell 30:758–786. http://kryten.mm.rpi.edu/SB_AS_CreativeSelf-DrivingCars_0323161130NY.pdf. (The URL here goes only to an uncorrected preprint)
20. Cope D (2005) Computer models of musical creativity. MIT Press, Cambridge, MA
21. Dennis L, Fisher M, Slavkovik M, Webster M (2016) Formal verification of ethical choices in autonomous systems. Robot Auton Syst 77:1–14. https://doi.org/10.1016/j.robot.2015.11.012
22. Luger G (2008) Artificial intelligence: structures and strategies for complex problem solving, 6th edn. Pearson, London, UK
23. Russell S, Norvig P (2020) Artificial intelligence: a modern approach, 4th edn. Pearson, New York, NY

The (Uncomputable!) Meaning of Ethically Charged Natural Language, for Robots, and Us, from Hypergraphical Inferential Semantics

Selmer Bringsjord, James Hendler, Naveen Sundar Govindarajulu, Rikhiya Ghosh, and Michael Giancola

Abstract The year is 2030. A two-young-child, two-parent household, the Rubensteins, owns and employs a state-of-the-art household robot, Rodney. With the parents out, the children ask Rodney to perform some action α that violates a Rubensteinian ethical principle P_R. Rodney replies: (s_1) "Doing that would be (morally) wrong, kids." The argument the children give Rodney in protest is that another household, the Müllers, also has a robot, Ralph; and the kids argue that *he* routinely performs α. As a matter of fact, Ralph's doing α violates no Müllerian ethical principle P_M. Ralph's response to the very same request from the children he tends is: (s_2) "Okay, doing that is (morally) fine, kids." What is the *meaning* of the utterances made by Rodney and Ralph? We answer this question by presenting and employing a novel, formal, inferential theory of meaning in natural language: *hypergraphical inferential semantics* ($\mathscr{HIS}$), which is in the general spirit of **proof-theoretic semantics**, which is in turn antithetical to Montagovian **model-theoretic semantics**. $\mathscr{HIS}$, applied even to sentences logically simpler than s_1 and s_2, implies that human-level natural language understanding (NLU) is Turing-uncomputable.

S. Bringsjord (✉)
Rensselaer Polytechnic Institute (RPI), Rensselaer AI & Reasoning (RAIR) Lab, Department of Computer Science, Department of Cognitive Science, Troy, NY, USA
e-mail: selmer.bringsjord@gmail.com

J. Hendler
Troy 12180, NY, USA
e-mail: hendler@cs.rpi.edu

N. S. Govindarajulu · R. Ghosh
Rensselaer Polytechnic Institute, Rensselaer AI & Reasoning Lab, Troy 12180, NY, USA

M. Giancola
Rensselaer Polytechnic Institute, Rensselaer AI & Reasoning Lab, Department of Computer Science, Department of Cognitive Science, Troy 12180, NY, USA
e-mail: mike.j.giancola@gmail.com

© The Author(s), under exclusive license to Springer Nature Switzerland AG 2023

M. I. A. Ferreira and M. O. Tokhi (eds.), *Towards Trustworthy Artificial Intelligent Systems*, Intelligent Systems, Control and Automation: Science and Engineering 102,
https://doi.org/10.1007/978-3-031-09823-9_11

1 Introduction

The year is 2030. A two-young-child, two-parent household, the Rubensteins, owns and employs a state-of-the-art household robot, Rodney. With the parents out, the children ask Rodney to perform some action α that violates a Rubensteinian ethical principle P_R. Rodney replies: (s_1) "Doing that would be (morally) wrong, kids." The rationale the children give Rodney is that another household, the Müllers, also has a robot, Ralph; and the kids argue that *he* routinely performs α. As a matter of fact, Ralph's doing α violates no Müllerian norm N_M. Ralph's response to the very same request from the children he tends is: (s_2) "Okay, doing that is (morally) fine, kids." We briefly explain herein how, given past work on our part, Rodney and Ralph would in general be engineered so as to respond in the (correct, for reasons explained) ways they do. But there is a separate issue, one that our prior work hasn't addressed; that issue is: What is the *meaning* of the utterances made by Rodney and Ralph? We answer this question by presenting and employing a novel, formal, inferential theory of meaning in natural language: *hypergraphical inferential semantics* ($\mathscr{HIS}$), which is in the general spirit of **proof-theoretic semantics**. HIS is based on a wholesale rejection of the dominant formal theory of the meaning of natural language: **model-theoretic semantics** (MTS), as seminally introduced by Montague [20]. We recommend that household robots (and *a fortiori* robots that frequently find themselves in morally charged situations, e.g. military robots) be engineered on the basis of the computational logics and corresponding procedures that underlie $\mathscr{HIS}$.

The remainder of our chapter unfolds in the following sequence. First, we present the case study involving robots Rodney and Ralph, and their respective families (Sect. 2). Next, in Sect. 3 we quickly explain how, given past work, Rodney and Ralph would in general be engineered. In Sect. 4 we very briefly summarize MTS, including—at least as the lead author sees things—some its fatal defects. The following section, Sect. 5, is a summary of proof-theoretic semantics for natural language, and a quick critique of this approach to meaning, in the form of today's state of the art. We then (Sect. 6) present (for the very first time in any archival venue) hypergraphical inferential semantics = $\mathscr{HIS}$, albeit briefly. Next, we apply $\mathscr{HIS}$ to the Rodney-Ralph case study (Sect. 7). Section 8 is devoted to the consideration of objections to what has come before. We then come to what may be the most impactful part of the present chapter: We show in Sect. 9 that the problem of determining the meaning of natural language such as s_1 and s_2 is not just challenging, and in fact not just possibly infeasible, but is in fact Turing-uncomputable. The chapter ends (Sect. 10) with a wrap-up, and an anticipatory look into the future regarding $\mathscr{HIS}$ both in general, and specifically in connection with machine ethics.

2 A Household-Robot Case Study

Rodney is a state-of-the-art English-speaking household robot of 2030, recently purchased by the Rubenstein family to help shop, cook, clean, and assist with various child-rearing tasks. Mr & Mrs Rubenstein are two 70-hours-a-week Manhattan corporate attorneys; they have active, exceptionally bright, and somewhat mischievous twins (Joel and Judith) who recently entered third grade in a nearby Upper-West-Side (secular) school: The Anderson School, for gifted and talented children (PS 334). The twins are also in Hebrew school two days a week, and have been since they started reading; both of their parents are modern-orthodox Jews, and are raising the twins in fastidious conformity with most of the tenets and practices embraced by adult members of this denomination.

Rodney has been tasked by the twins' parents at sunrise with grocery shopping and making dinner for the twins at 6pm. Their parents, who now leave for work, will be toiling late at their offices. Judith, perhaps testing Rodney a bit later at breakfast: "Rodney, can we please have Lobster Newberg for dinner today? My Christian friend at school says it's delicious, and their robot made it for them!" What, for Rodney and Ralph, resp., is the meaning of following three normative sentences?

---

s_1' It is morally forbidden for Judith and Joel to have lobster for dinner, and for their robot Rodney to cook such lobster.

s_2' It's morally permissible for some Anderson students to have lobster for dinner, and for their robot Ralph to make such a meal.

s_3 It is wrong for Judith and Joel to plan to have lobster for dinner, yet permissible for them to entertain having such a meal.

The third sentence here involves the moral status of mental acts, and is beyond the scope of the present chapter, the chief purpose of which is to introduce $\mathscr{HIS}$ in connection with both norms, and human-robot interaction, and to reveal that natural language understanding (NLU) in $\mathscr{HIS}$ is Turing-uncomputable. But we do herein answer the question about the first two of the sentences here, by presenting and employing a novel semantics for such modal propositions in the general spirit of proof-theoretic semantics. Our approach covers the semantics of natural-language sentences such as those listed above. (Needless to say, any headway we make would be applicable to the meaning of norms associated with roughly parallel propositions that confront robots employed by Hindus, Christians, and Muslims, but also—since the core challenge isn't restricted to divine-command ethical theories and codes—atheists, utilitarians, and so on.

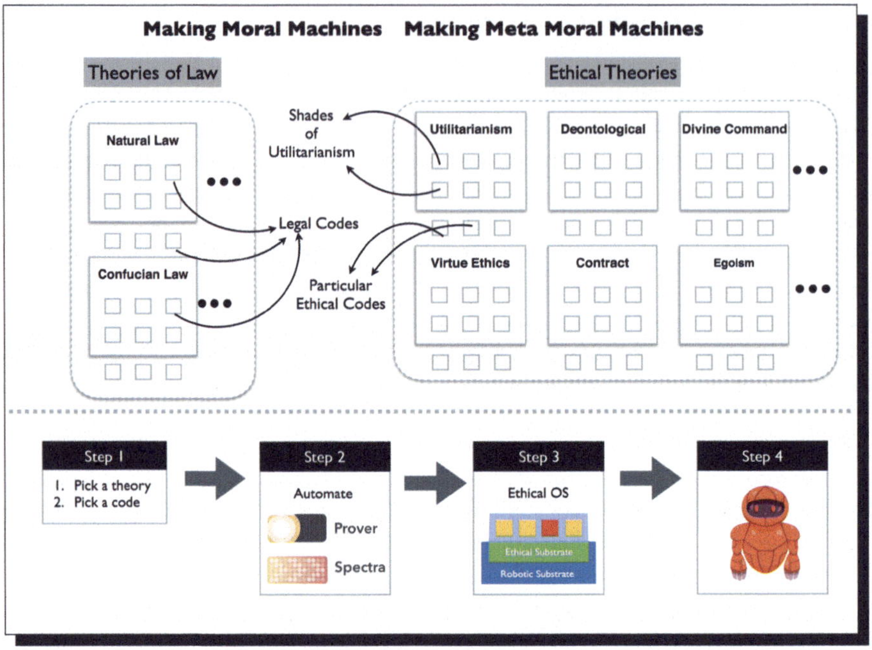

Fig. 1 The four steps in making ethically correct machines

3 Prior Framework for Engineering of Robots Rodney and Ralph

Work by Bringsjord and Govindarajulu (and some collaborators with them) through the years that has been devoted to the science and engineering needed to achieve ethically correct robots has been firmly logicist in nature. Overall this effort abides by the "Four-Step" approach shown from a high-level point of view pictorially in Fig. 1. We now quickly summarize these four steps, in sequence, so that the reader will be in position to understand how $\mathscr{HIS}$ and NLP built upon it fits quite naturally with prior work.

The first step is selecting an ethical theory from a family thereof. We do not want to advance a framework that requires one to commit to any particular ethical theory or even to families of theories. In the case study at hand, we of course assume that Rodney's family and Ralph's family each affirm different (indeed inconsistent) ethical theories (even if only implicitly).

So, assume that we have a family of ethical theories $\mathscr{E}$ of interest. We assume that, minimally, any ethical theory $\mathcal{E} \in \mathscr{E}$ obligates or permits (i.e. sanctions) a set of situations or actions Π and forbids a set of other situations or actions Υ. When these situations become particular, we are dealing with a moral code X based on the theory $\mathcal{E}$; such codes are by definition domain-dependent. For example, both our families, the Rubeinsteins and the Müllers, have particular ethical codes governing diet.

Abstractly, assume that we have a formal system $\mathcal{F} = \langle \mathcal{L}, \mathcal{I} \rangle$ composed of a language $\mathcal{L}$ and a system of inference schemata (or a proof theory/argument theory) $\mathcal{I}$. The particular formal system, a so-called *cognitive calculus*, that has been much used in the past for modeling and simulating ethical reasoning and decision-making in AIs and robots is $\mathcal{DCEC}$; see e.g. [13]. One non-negotiable *sine qua non* for the kind of calculus we need, one (as will be seen), directly relevant to determining the meaning of the two key sentences s_1' and s_2' from our case study, is that quantified formulae containing the deontic modal operator **O**, for 'is obligatory,' must be available.

The second of The Four Steps is to automate the generation of proofs of (un-)ethical behavior so that the reasoning can be utilized and acted upon by autonomous robots. We use ShadowProver [16], an automated reasoning system (among other things) tailor-made for use of $\mathcal{DCEC}$.

The third step is to integrate this ethical reasoning system into an autonomous robot's operating system, something that, longer term, we would insist upon for both Rodney and Ralph, were these robots of our own design. For reasons explained in [15], there are basically two possible approaches to this (see Fig. 2). In the first, only "obviously" dangerous AI modules are restricted with ethical reasoning safeguards *installed above the OS*. In the second approach, and by our lights highly preferable one, all AI modules must be brought down to the robotic substrate (the percepts and actuators which enable the robot to interact with its environment) through an "Ethical Substrate" tied to the OS). The advantage of the first approach is speed: modules which are not inhibited by an ethical safeguard are able to directly manipulate the robot. However, this option also allows for the possibility that those AI modules deemed "not dangerous" may end up making a decision which leads the robot to act unethically. Only in the second option is ethical behavior guaranteed.[1]

In the fourth and final step, we implement, and thereby arrive at a moral machine, in the real world.

4 Montagovian/Model-Theoretic Approach to Meaning, Rejected

At least until today, by far the dominant approach to formally pinning down the meaning of natural language is **model-theoretic semantics** (MTS), seminally introduced and—at least to an impressive degree—specified by Montague [20]. In this section we quickly encapsulate MTS, and then explain why it must be rejected in light of its being plagued by a series of fatal defects.

[1] That is, ethical behavior relative to some ethical theory, and code selected therefrom.

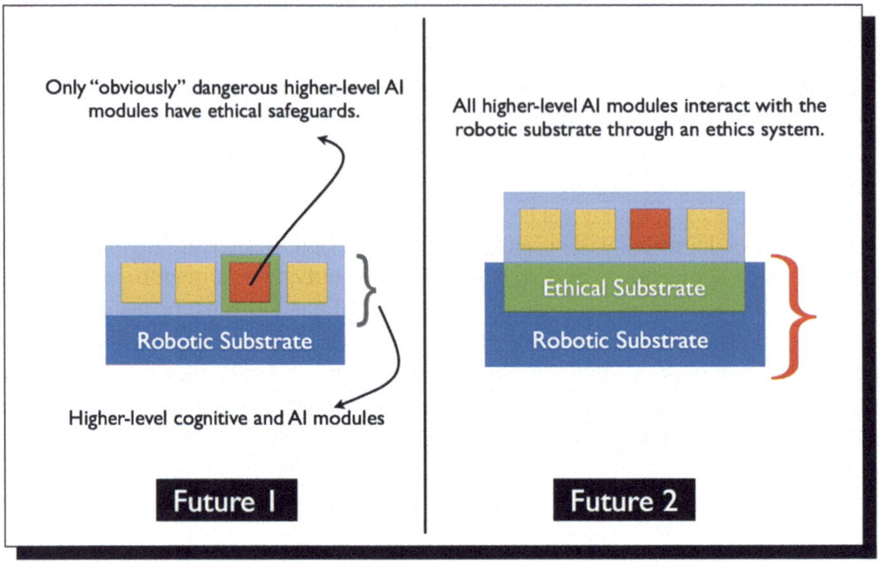

Fig. 2 Two Futures—with and without an ethical substrate. *Higher-level modules are vulnerable to tampering. The ethical substrate protects the robotics substrate from rogue modules*. Figure from [15]

4.1 MTS in Summary

We don't pretend that we can do justice to MTS here; but we say a few words, and hope they are helpful: MTS, in the case of formal logic, as we've already indicated, is Tarskian, and says that the meaning of formulae consist in the conditions for their being TRUE on interpretations, compositionally calculated. For instance, for any interpretation $\mathcal{I}$ whose domain of quantification includes no red things, the formula

$$\forall x(Red(x) \rightarrow Happy(x))$$

will be TRUE.[2]

What about MTS not for logic, but natural language? Well, actually, at least when it comes to the meaning of sentences in e.g. English, meaning is delivered in a manner akin to how it works for formulae. For instance, at least on the brand of MTS advanced by Montague himself, English nouns, verbs, and adjectives become relation symbols in a formal (logical) language, and the meanings of these relations are just suitable tuples of objects in some domain for some interpretation.

[2] Note that the relation symbol *Red* doesn't appear anywhere above in the present paper. That is as desired, because no matter what relation symbol *R* is used in a simple quantified conditional of the form we use here, if the domain of quantification has no non-empty class to which this *R* is mapped, the formula has a meaning of TRUE.

4.2 Two Fatal Defects in Model-Theoretic Semantics

At least according to Bringsjord, MTS is fatally flawed. Here, given space constraints, and given as well that the focus is on the presentation of $\mathscr{HIS}$ for purposes of handling the meaning of norms for robots like Rodney and Ralph, only two such defects are mentioned, and both defects are at best synoptically communicated. Here is the pair:

1. *Everything Grounds Out in Proof/Argument.* MTS has its roots in what is known today in mathematical logic as *model theory*, which was, if not outright invented by Tarski (Montague's advisor), then at least brought to a respectable level of formality by him. In order to begin to see that model theory and the "meaning" it assigns to formulae distills to meaning cast in terms of inference, and specifically proofs or arguments, consider what the meaning of a simple material conditional

$$c := \phi \rightarrow \psi$$

is according to the model theory for first-order logic, where ϕ and ψ are wffs in the standard formal language of this logic.[3] We have that the meaning of c is either TRUE or FALSE; where the former case holds if and only if—and here we quote from relevant mathematical-logic coverage of model theory, e.g. [9]—"If (interpretation) $\mathcal{I}$ satisfies ϕ, then $\mathcal{I}$ satisfies ψ as well." So we have exchanged a formal material conditional for ... another conditional, one that is a mix of the formal and informal. Now what determines whether this hybrid conditional holds? Well, we need a meta-logical proof that this hybrid conditional is true (by e.g. supposing that its antecedent holds, and proving that based on this supposition the consequent holds as well). This is to say that the meaning of even a dirt-simple material conditional ends up being a matter of proof carried out at the meta-logical level, over the formal elements of model theory![4]

[3] Please note that MTS is certainly up to the challenge of producing meaning for things much, much more robust than material conditionals, but the point here is that *even* in the case of something as simple as a material conditional treated by model theory in standard, elementary, classical mathematical logic, meaning reduces to meaning in terms of inference. In addition, we are certainly aware of the obvious fact that no one working on formal semantics believes that material implication is a good representation of natural-language conditionals. But this fact is orthogonal to the point we are making here: that, again, meaning initially taken to be model-theoretic eventuates in meaning that is inferential in nature.

[4] The disappearance of the mirage that meaning can be at the level of models/model theory carries over *mutatis mutandis* directly to English. An instance of c in English might for instance be

$$c' := \text{If Johnny helped, Olaf did too.}$$

The meaning of c' can't be in any model-theoretic basis. If it is said that c' is TRUE because it holds in some particular interpretation $\mathcal{I}^*$, one has only ask why this is the case. The only cogent response is to supply the relevant formal machinery and associated inferences (e.g. that as a matter of fact $\mathcal{I}^*$ renders 'Johnny helped' TRUE).

2. *Possible Worlds are at Best Merely Metaphorical, and at Worst Provably Incoherent.* While Kripke gets credit for seminally working out so-called "possible-world semantics" formally, Leibniz had an intuitive concept of, and wrote about, possible worlds. But what *is* a possible world? This question is distressingly hard to answer, for everyone—to this day. But the situation is actually worse, because some answers that were confidently proffered on the strength of basic set theory and consistency turned out to be *provably* incoherent [1]. Common practice is to just take the concept of a possible world as an unanalyzable primitive, but then the obvious question is: How is it that meaning gets explicated in terms that, by definition, are not assigned a meaning?

5 Proof-Theoretic Semantics (PTS)

5.1 The Basic Idea

The basic idea behind PTS, at least in its modern form and officially speaking, originates with Gentzen [12], commonly regarded to be the inventor of natural deduction, and is not unfairly encapsulated thus: The meaning of elements of a proof, say for instance a constant a in a proof π, consists in the instantiation of inference schemata in π to introduce a. Once one grasps the basic insights of Gentzen, and the subsequent extensions of Prawitz [22], and combines these insights with what we have shown above about the unstoppable grounding out of model-theoretic truth/falsity in proof, it isn't long before those new to PTS, but well-versed in formal logic and mathematics, see at least the possibility of claiming that *all* meaning, at least for coherent declarative content, consists in the position of this content within proofs. Interestingly enough, professional mathematicians deal in proof top to bottom and beginning to end, and have for millennia, but know next to nothing about model theory in any form. This is often taken by advocates of PTS to be a tell-tale phenomenon; it certainly is by the lead author of the present paper. The body of technical literature on PTS is now vast, and we can say no more in terms of an overview, but (1) we direct interested readers to this starting place: [25][5]; and when we below speak about hypergraphical natural deduction and $\mathscr{HJS}$ itself, the reader will learn and understand more about PTS.

5.2 But What About PTS for Natural Language?

An impressive advance in proof-theoretic semantics for natural language has been achieved in Part II of [11]. (We do not have the space here to recount what is done

[5] More philosophically inclined readers should without question read Dummet's [8] remarkable attempt to erect a theory of meaning along the PTS line.

in this work.) Unfortunately, as impressive as this book is, there are some serious inadequacies, especially in the context of our case study regarding robots Rodney and Ralph. Here are two such problems:

- *Deduction/Proofs Only*. As even readers new to formal semantics doubtless imagined when reading for the first time about MTS vs. PTS above, given that the 'P' in 'PTS' is for 'proof,' meaning on this approach must ultimately be cashed out as proofs and their constituents and use. But this dooms PTS at the outset, for the simple reason that most reasoning engaged in, explicitly and implicitly, by humans, is non-deductive in nature. Consider for instance what the children say to Rodney in an attempt to persuade him to prepare Lobster Newberg. They give him not a proof, but an argument (one based on a failed analogy between two households).
- *No Operators for Ethical and Cognitive Phenomena*. As we have seen, we need to be able to speak about what is morally *obligatory*, *permissible*, and *forbidden* (minimally), and we certainly need to be able to speak about what agents believe and know (including about what other agents believe and know). But these needs, in formal logic, for reasons that can be expressed in the form of telling proofs [4], call for modal logic (in particular, resp., deontic modal logic & epistemic logic). Unfortunately, Francez [11] works with formal machinery that is devoid of modal operators, and for this reason sentences like our s'_1 and s'_2 can't be handled by his logical machinery.

6 Hypergraphical Inferential Semantics ($\mathscr{HIS}$)

In this section we provide a brief overview of a novel formal theory of meaning, Hypergraphical Inferential Semantics ($\mathscr{HIS}$), which is inspired and guided by hypergraphical reasoning—and also of course by the two inadequacies cited in the previous section. The overview proceeds as follows. We first (Sect. 6.1) convey, intuitively, the apparent brute fact that the meaning of natural language hinges on inferential context. Next, we give a very brief explanation of hypergraphical natural deduction (Sect. 6.2). We end the present section with an example, one in which the seemingly humble sentence 'Emma helped' is given agent-indexed meaning on $\mathscr{HIS}$ (Sect. 6.3).

6.1 Intuitive Kernel of $\mathscr{HIS}$ via Buffalo-Buffalo-...

Consider this sentence:

(2) Buffalo buffalo buffalo.

What does (2) mean? You don't know. Upon some reflection, though, you will certainly find yourself entertaining some possibilities. Which of these possible meanings is what (2) means? For example, does (2) mean nothing; i.e. is it just three occurrences of the word that denotes the species of the animal *B. bison*, and nothing more? Maybe; but then again maybe not. Suppose we trustworthily tell you that (2) has been uttered somewhat slowly by Smith as he thinks back wistfully to a time when vast numbers of buffalo roamed proudly across portions of North America, before their rapid decline in the 1800s.s. In this case, (2), given what we have just told you and inferences made therefron, means something like:

(2_m) Smith believes that impressive and even glorious must have been the status of the mighty buffalo across the great midwestern plains and the foothills of the Rockies before heartlessly preyed upon by man!

One might wonder why Smith is thinking back to the "glory days" of American bison. There could of course be any number of reasons for his contemplation. For instance, suppose that Jones said the following, just before Smith utters (2):

(1) They can reach twelve feet in length, weigh over 2,000 pounds, and imagine horde upon horde of them in the wild, before the great slaughter, thundering sometimes in full, 40-miles-per-hour stampedes beneath the peaks of the Tetons, nothing to fear.

Given (1) beforehand, (2)'s meaning (2_m) is quite plausible—because there exists an obvious argument (which we don't detail) from (1) and other declarative information to (2_m). A bit more precisely, given (1), and background propositions about human psychology, aesthetics, and so on, that (2) means (2_m) is just to say that (2_m) is the conclusion of an argument. However, suppose instead that (1) was never uttered, but rather that our Smith is reading an article by an august naturalist in which this author claims that

($1'$) Some buffalo in Buffalo hoodwink other buffalo.

and that upon taking this in, Smith murmurs a "Hmm" and a "So" to himself, and then says (2). The meaning of (2) is now nothing at all in the vicinity of (2_m). Instead, the meaning of (2) is ($1'$) = ($1'_m$) itself, and Smith has simply affirmed an argument from what he has read to the pinning down of meaning.[6]

The moral of all this talk of buffalo should be clear. It's that the meaning of natural-language sentences (at least frequently, and perhaps always) consists in their being within arguments (or, in more rigorous situations, proofs).

6.2 *Hypergraphical Natural Deduction*

We assume readers to be familiar with basic graph theory, and to therefore be acquainted with *directed hypergraphs*. $\mathscr{HIS}$ takes the meaning of a natural-language sentence s to consist in the location of s within a (usually vast, in "real life")

[6] It's interesting to note that any debate about the meaning of (2) in the contexts we have laid down will just end up providing further evidence for the view that meaning is inferential (since debate is, if anything, inference-based).

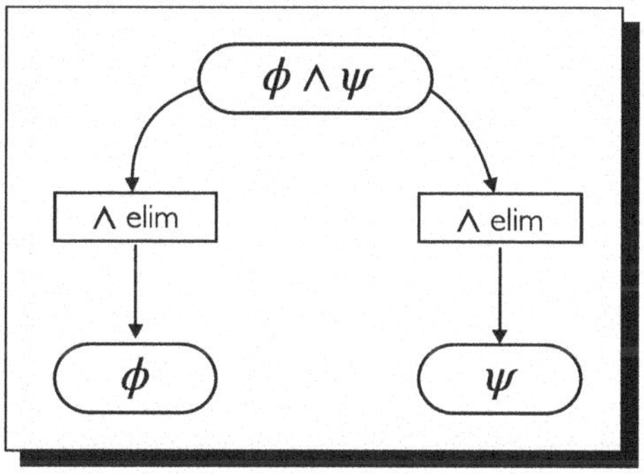

Fig. 3 Hypergraphical $\wedge$ `elimination`

directed hypergraph that specifies interacting arguments and proofs. These graphs are dynamic, since human reasoning, as long noted in AI, is nonmonotonic; but in the present paper we ignore dynamism and worry only about meaning *at a particular time*. We also assume readers to be familiar with basic natural deduction in its standard forms. But now, what about *hypergraphical* natural deduction? And indeed, more broadly, hypergraphical reasoning? The basic concept of such formalisms date back to [7], but ignoring for economy here the development of these ideas through time, and the implemented proof- and argument-construction environments available today, we greatly simplify and first draw your attention to an interesting quote from Schroeder-Heister [25], who writes:

> One could try to develop an appropriate intuition by arguing that reasoning towards multiple conclusions delineates the area in which truth lies rather than establishing a single proposition as true. However, this intuition is hard to maintain and cannot be formally captured without serious difficulties. Philosophical approaches such as those by Shoesmith and Smiley (1978) and proof-theoretic approaches such as proof-nets (see Girard, 1987; Di Cosmo and Miller, 2010) are attempts in this direction. ([25], Sect. 3.5)

What is here declared to be just an "attempt" is made perfectly concrete in hypergraphical reasoning. Consider the pair of Figs. 3 and 4, to which we draw your attention now.

Notice here that reasoning isn't linear: conclusions drawn as a result of inferences are in no way done one at a time in step-by-step fashion in a single list of formulae. On the contrary, what Schroeder-Heister indicates has "serious difficulties" has absolutely none at all. Multiple conclusions of ϕ and ψ in Fig. 3 happens simultaneously in the directed hypergraph shown there. And of course in Fig. 4, two premises, ϕ on the left and ψ on the right, lead *at once* in the graph to the conjunction.

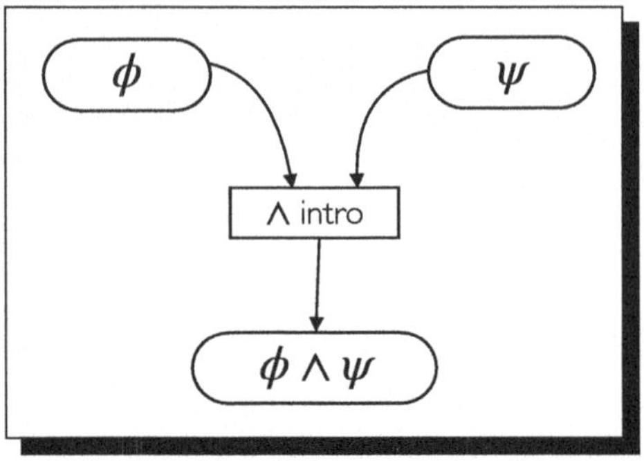

Fig. 4 Hypergraphical $\land$ `introduction`

6.3 An Example: The Meaning of 'Emma Helped'

What is the meaning of the two-word English sentence that immediately follows?

s Emma helped.

Given the foregoing, the reader knows that the initial, provisional answer advanced by at least the lead author of the present paper is: "Well, it depends on inferential context." Of course, this is a programmatic answer, not a genuinely informative one. Let us then set some context, by stipulating that two pieces of declarative information are givens for the agent who reads or hears s, to wit:

—

G1 The following three propositions are either all true, or all false.

1. If Billy helped, Doreen helped.
2. If Doreen helped, Frank helped.
3. If Frank helped, Emma helped.

G2 Billy helped.

—

Now, in the context composed by givens G1 and G2, and specifically assuming that first G1 and G2 are assimilated, what is the meaning of s? The answer is still "It depends." The reason is that the meaning of s for a given agent $\mathfrak{a}$ who has taken in first both G1 and G2, and then s, will depend upon the hypergraphical natural deduction that has now formed inside s. For a rational agent, the meaning of s will correspond to the hypergraphical proof shown in Fig. 5. Such an agent will be able to confidently report that given G1 and G2, s is true.

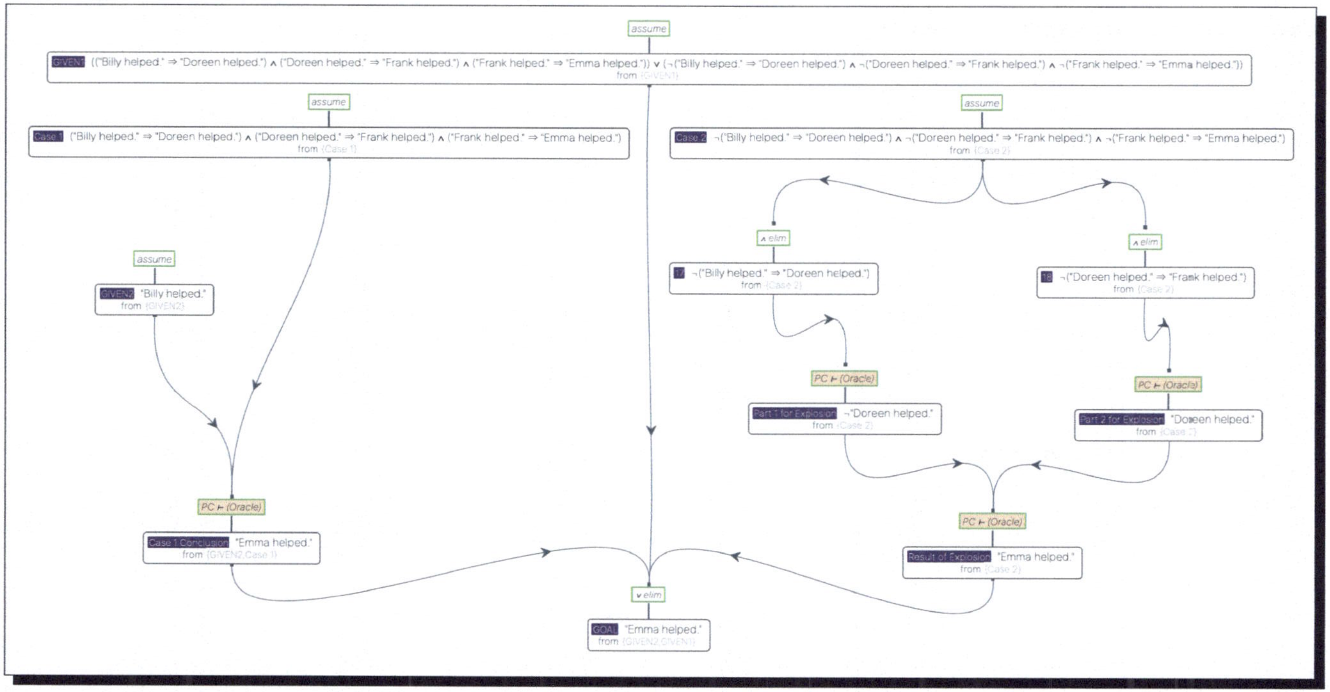

Fig. 5 Semi-automated proof of 'Emma helped' From G1 and G2. *This is the meaning,* $\mathscr{HIS}$ *of 'Emma helped' for the agent who understands that, as a matter of fact, assuming both G1 and G2, Emma did help*

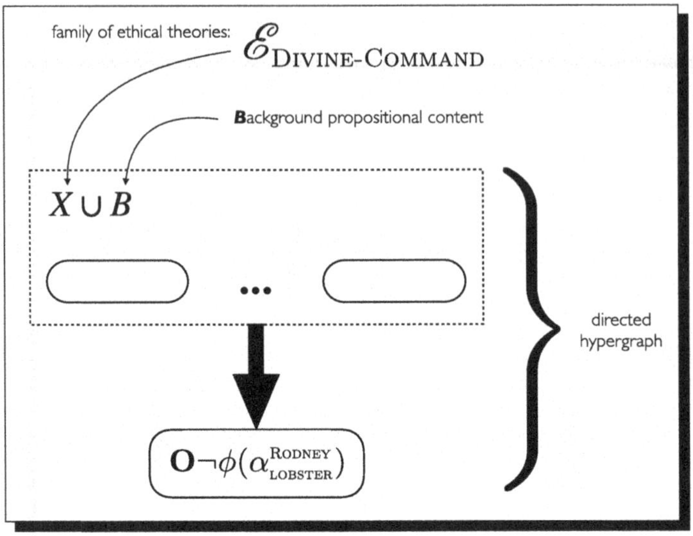

Fig. 6 The meaning of the sentence s_1' according to $\mathscr{HIS}$

7 Applying $\mathscr{HIS}$ to the Robot Case Study

It should be rather clear to the reader at this point what the meaning of our featured sentences are. That meaning consists in a directed hypergraph, indexed to a particular agent, and anchored in the elements of the Four-Step Process shown in Figure 1. What elements? First, the relevant family of ethical theories is that of *divine-command* sort.[7] From this family a particular theory associated with the relevant sort of Judaism is selected, and from that is in turn selected a particular ethical code X. When this code is combined with background B declarative information, a proof, or at least an argument, for the formula that expresses s_1' can be inferred. This formula is

$$\mathbf{O}\neg\phi(\alpha^{\text{RODNEY}}_{\text{LOBSTER}}),$$

where $\phi(\_, \_, \ldots, \_)$ is an open formula with "placeholders" for the relevant parameters in the case study. For a depiction of the overall situation, and the meaning of s_1' as a hypergraph, see Fig. 6.

Figure 7 shows a Python program that calls ShadowProver, which is able to find a fully-automated proof of sentence s_1' in under a second.

But what about Ralph? Why is it ethically permissible for him to whip up Lobster Newberg for the children he tends to? More to the matters at hand, what is the meaning

[7] In the context of machine ethics, the formalization of this family is explored in [5]. A seminal treatment, from the point of view of analytic philosophy and formal logic, of this family of ethical theories, in particular the sub-family associated with Judaism and Christianity, is given by Quinn [23].

```
import sys
sys.path.insert(1, '/pylibs/interface')
import interface

from time import time


if __name__ == "__main__":
    assumptions = [ "(Knows! rodney t (CulinaryCode rubensteins))",

                    "(Knows! rodney t (implies (CulinaryCode rubensteins) \
                                      (forall [?f] (iff (ForbiddenFood ?f) \
                                                  (not (Kosher ?f))))))",

                    "(Knows! rodney t (not (Kosher lobster)))",

                    "(Knows! rodney t (forall [?f] (implies (ForbiddenFood ?f) \
                                              (Ought! rodney t (ForbiddenFood ?f) \
                                                (not (happens (action (rodney (cook ?f))) t))))))"
                  ]

    goal = "(Ought! rodney t (ForbiddenFood lobster) \
            (not (happens (action (rodney (cook lobster))) t)))"

    start = time()
    print(interface.prove(assumptions,goal))
    stop  = time()

    print("Time: " + str(stop - start) + " seconds.")
```

Fig. 7 An automated proof of the sentence s_1' in accordance with $\mathscr{HIS}$, found. (*Needless to say, this is inteneded to convey but the gist of what in its full, real-world version would be rather elaborate, since it would need to be aligned and integrated with the rigorous and subtle theological reasoning of relevant humans (e.g., rabbis). In addition, an argument is much more likely to ultimately be in play, rather than a proof; and probability/likelihood, rather than only classical bivalence, would inevitably be in play as well.*)

of sentence s_2'? As alert readers can doubtless surmise, we have pretty much a direct parallel to what we've already seen in the case of Rodney—save for some obvious differences. First, of course the formula that is at the end of the relevant directed hypergraph expresses that the relevant culinary action is ethically permissible for him; this formula is:

$$\neg\mathbf{O}\neg\phi(\alpha^{\text{RALPH}}_{\text{LOBSTER}}),$$

The overall situation w.r.t. Ralph is shown in Fig. 8, and the program generating the corresponding proof (via ShadowProver) in Fig. 9. Again, ShadowProver found a fully-automated proof in under a second.

8 Questions/Objections and Replies

We here consider some questions and objections, and reply to each, in short.

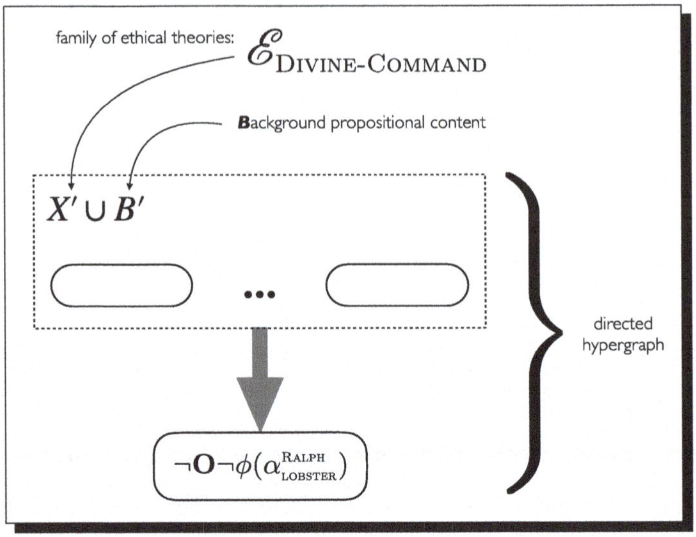

Fig. 8 The meaning of the sentence s_2' according to $\mathscr{HIS}$

```
import sys
sys.path.insert(1, '/pylibs/interface')
import interface

from time import time


if __name__ == "__main__":
    assumptions = [ "(Knows! ralph t (CulinaryCode mullers))",

                    "(Knows! ralph t (implies (CulinaryCode mullers) \
                                  (forall [?f] (iff (ForbiddenFood ?f) \
                                           (and (DuringLent t) (IsMeat ?f))))))",

                    "(Knows! ralph t (not (IsMeat lobster)))",

                    "(Knows! ralph t (forall [?f] (implies (not (ForbiddenFood ?f)) \
                                         (not (Ought! ralph t (not (ForbiddenFood ?f))  \
                                           (not (happens (action (ralph (cook ?f))) t)))))))"
                  ]

    goal = "(not (Ought! ralph t (not (ForbiddenFood lobster)) \
               (not (happens (action (ralph (cook lobster))) t))))"

    start = time()
    print(interface.prove(assumptions,goal))
    stop  = time()

    print("Time: " + str(stop - start) + " seconds.")
```

Fig. 9 An automated proof of the sentence s_2' according to $\mathscr{HIS}$

8.1 *Objection 1: Redundancies & Irrelevancies*

"Generally, automatically-derived proofs are very complex structures, in no small part because they can contain a lot of redundancies and irrelevant steps. How do you deal with this when seeking to model communication between humans?"

True enough, an arbitrarily discovered proof in response to a query as to whether some formula ϕ can be derived from some starting collection Φ of formulae may not at all be streamlined, let alone minimal. But all that $\mathscr{HIS}$ posits is the existence of a proof or argument from some relevant Φ to some relevant ϕ that corresponds to what is happening in given rational communication between humans. No reason is given here to think that this proof or argument cannot be found computationally, for an NLU system.

8.2 Objection 2: Termination

"Your formalism is not completely specified but does seems rather powerful. Can you guarantee that the proof procedure always terminates?"

Yes. But there is a caveat! By Church's Theorem, the *Entscheidungsproblem* is only semi-decidable; i.e., once we hit first-order logic, theoremhood is at best semi-decidable; and, as we have explained, we are well beyond first-order logic when the meaning of natural language is given inferentially. Nonetheless, termination for a Σ_1 process can be guaranteed to terminate by use of a timer cutoff, a technique no different than what's available on a contemporary smartphone. We refer to a countdown timer, which can be engaged in such a way that some action is performed when the countdown ends. In parallel, we can simply pick some amount of time beyond which search is prohibited, and produce whatever the result is at the moment that time expires.

8.3 Objection 3: How are Divine-command Theories and Codes Formalized?

"When it comes to deriving norms, your presentation is insufficient, in my opinion. It would be really interesting to understand how the ethical code of the divine-command theory is formalized. And your presentation does not allow the reader to understand how the obligations are actually derived."

Undeniably, this objection opens up deep issues that can't possibly be treated in the present venue, which is after all chiefly intended to present $\mathscr{HIS}$. The reader will need to turn to some of our writings that are focused on machine ethics, not formal semantics, to obtain answers. For instance, the Doctrine of Double Effect is an ethical principle that is at the code level, not at the ethical-theory level; and the reader can consult [13] to see in detail how this principle is formalized. Put roughly, some proposition is at the ethical-theory level if it is a biconditional that provides the necessary and sufficient conditions according to which an action is obligatory (or forbidden, supererogatory, etc.) in general. In the case of divine-command ethical theories, one such theory is given in [23]; but many other divine-command theories are possible. Our robot Rodney would be working under one such theory. As to the code level, prohibitions regarding diet, which are of course central to the case study with which we began, are at that level. Code-level propositions pertain to particular actions or action classes, and their moral status; ethical-theory-level propositions are about how to define obligation and other concepts in general. Finally along this general line, it's important to understand that *any* family of theories can be used in our Four-Step approach, including consequentialist theories (including types of utilitarianism). Nothing in what we have said above precludes using $\mathscr{HIS}$ to systematize the meaning of theory-level or code-level propositions.

8.4 Objection 4: OS-Level Ethical Controls Mysterious

"I'm not sure what is meant by 'installed above the OS,' and 'bring all AI modules down to the percepts and actuators which enable the robot to interact with its environment,' and here I quote what you have said above."

Addressing such concerns is out of scope here, and we must accordingly direct the reader to [14], since the main purpose of the present paper, again, is to explain how the meaning of moral obligations for machines, specifically household robots, can be determined by our general approach.

9 Natural Language Understanding Is Provably Uncomputable

As promised above, we briefly show in this, the penultimate section of our chapter, that human-level natural language understanding (NLU), when construed in general as the problem of receiving some arbitrary natural language S at this level, along with associated content, and producing in response the meaning of S per $\mathscr{HIS}$, is Turing-uncomputable (hereafter just 'uncomputable' *simpliciter*).[8] For simplicity, but with no loss of generality, assume that S is composed of a finite set of sentences $s_1, s_2, \ldots, s_k$. In fact, still without loss of generality, and perhaps surprisingly to some readers, we can restrict our attention to the case in which S is composed only of buffalo sentences (= b-sentences); these are of course the type of sentences we discussed earlier in Sect. 6.1. Now, however, we shall need to get more precise about b-sentences, and we start to do this by briefly considering formal grammars for such sentences.

We now fix our first exceedingly simple grammar $\mathcal{G}_b^{\mathscr{L}_{pc}}$. The purpose of the subscript here is obvious; the superscript indicates that the grammar in question is at

[8] Of course, there are an infinite number of accounts of the meaning of human-level natural language on which arriving at meaning becomes not only Turing-computable, but polynomial/**P**. The point in the present section, expressed by the theorem below, is that *if* it's correct that the meaning of some natural language in the human case is inferential in nature as per $\mathscr{HIS}$, *then* the problem of producing meaning from relevant input is uncomputable.

the level of the propositional calculus. In this grammar, for a noun to denote the U.S. city of Buffalo, we use `Buffalo1`, and as a noun to denote the animals in question we employ `buffalo2` and—when majuscule is needed for the start of a sentence—`Buffalo2` as well. We additionally have for the verb in question `buffalo` (understood here, in keeping with standard English dictionaries, as "to intimidate by a display of power"), and here too we can if needed avail ourselves of the uppercase variant `Buffalo`.[9] For economy, we don't specify the grammar in BNF form, but such a specification should be obvious to the reader, and easily obtainable therefore.

Now let's move further toward establishing the theorem that the problem of arriving at meaning for an arbitrary b-sentence is uncomputable. To do this, note first that the following informal b-sentence is ambiguous:

s_3 *Buffalo buffalo buffalo buffalo.*

As to the context, we stipulate that it—for reasons beyond scope here—logically implies that the first word is a reference to the city of Buffalo that is very near Niagara Falls, the second to the animals in question, the third to our one and only verb, and the fourth and final word another reference to the animals. In our first formal grammar, then, sentence s_3 is apparently disambiguated as:

s_3' `Buffalo1 buffalo2 buffalo buffalo2.`

But note that, as a matter of fact, the ambiguity isn't resolved in the least here. Upon reflection, it should be clear why. What does (s_3') mean? Does it mean that *all* buffalo in the city of Buffalo buffalo *all* buffaloes? Or does it mean that *some* buffalo in the city of Buffalo buffalo *all* buffaloes? Clearly, even with only classical quantification in play implicitly, s_3/s_3' is ambiguous between four distinct candidate permutations. Let's then expand our formal grammar by moving to first-order logic = $\mathscr{L}_1$; we thus have—following the notation we have introduced—$\mathcal{G}_b^{\mathscr{L}_1}$, or simply $\mathcal{G}_b^1$. In this grammar, we allow `all` and `some`, with a direct match between these words and the two quantifiers $\forall$ and $\exists$ of $\mathscr{L}_1$. (Hence `some` is interpreted as "at least one.") Given this, here's one possible genuine disambiguation for s_3 and s_3':

s_3'' `All Buffalo1 buffalo2 buffalo all buffalo2.`

And here is the representation of s_3'' in first-order logic itself (with obvious use of abbreviatory relation symbols):

$\sigma_{(b1')}$ $\forall x[(B_1 x \land B_2 x) \rightarrow \forall y(B_2 y \rightarrow Bxy)]$

The reader should note that along this line, under the umbrella of $\mathscr{HIS}$ and our formalization of what NLU is, we quickly run into formulae that are not satisfied by any interpretation with a finite domain. The quickest way to explicitly see this is simply to note that, when it comes to even the grammar $\mathcal{G}_b^{\mathscr{L}_1}$ with a trivial expansion, we have sets of formulae in this category, for consider:

---

[9] Generally such a need only arises when we admit b-sentences that are *imperative* in nature (as e.g. in the command to a buffalo animal (or animal*s*) that it intimidate by a display of power: "Buffalo buffalo!"). We focus exclusively on *declarative* buffalo sentences in the present section/chapter.

extensional (classical):	$\mathcal{G}_b^0$	$\mathcal{G}_b^1$	$\mathcal{G}_b^2$	$\mathcal{G}_b^3$	...
extensional (non-classical):	$\mathcal{G}_b^0$	$\mathcal{G}_b{}'$	$\mathcal{G}_b{}''$	$\mathcal{G}_b'''$	...
intensional (classical):	$\omega\mathcal{G}_b^0$	$\omega\mathcal{G}_b^1$	$\omega\mathcal{G}_b^2$	$\omega\mathcal{G}_b^3$	...

Fig. 10 Part of the infinte array that composes an infinite family of buffalo grammars. (*To visualize this as an infinite tree, imagine that the array is rotated clockwise 45 degrees.)*

- `No buffalo buffalos itself.`
- `If a buffalo-1 buffalos a buffalo, and the buffaloed buffalo buffalos another buffalo-3, buffalo-1 buffalos buffalo-3.`
- `Every buffalo buffaloes some buffalo.`

This trio, when represented in first-order logic, cannot be satisfied by a finite model, since it's an isomorph of a well-known example from Kleene [18].

At this juncture we point out that with a base lexicon that is minuscule, what we are seeing is nonetheless the beginning of a progression out from this lexicon to a vast family of grammars. We don't have the space in the present chapter to define this family, but rest content with pointing out that it can be viewed as an infinite array that has increasingly complex languages appearing as the array builds out to the reader's right; see Fig. 10. In this figure, 'ω' indicates some collection of modal operators; for more along this line of abstracting to modal operators of any sort, see [3].

Very well. Now what of the theorem we are seeking? Given the foregoing in the present section, and given as well how $\mathscr{HIS}$ has been defined, we simply note that to determine the meaning of S it must specifically be determined whether, from a set Σ of formulae in the relevant formal language for the relevant formal logic, one or more formulae $\sigma \in \Sigma$ is such that σ is provable from relevant background content conjoined with S itself. But this then enables us to easily establish what we are seeking[10]:

[10] The proof here exploits a connection to Church's Theorem. Surely there must be prior arguments made for the uncomputability of at least some aspects of human natural-language "computing," ones that rely on other established negative theorems in recursion theory (such as the Post Correspondence Problem?)—but the first author is currently unaware of any, despite considerable digging.

Theorem: Uncomputability of Meaning (= UMT)

Theorem: Let s be some arbitrary grammatically correct sentence in a human-level natural language, and let σ_s^i be a representation of s in $\mathscr{L}_1$ from among an at-most countably infinite number of such representations. Then the meaning of s, by $\mathscr{HIS}$, i.e. some proof π or argument α, is uncomputable.

Proof: Suppose in particular that $s \in \mathcal{G}_b^1$, that $\mathbf{B}^s$ is the background propositional content for s, that X^s is specific, contextual information, and that for *reductio* the meaning of s is computable. Then for some particular k, whether

$$\mathbf{B}^s \cup X^s \vdash_{\pi/\alpha} \sigma_s^k$$

holds is computable. But this contradicts Church's Theorem. ■

9.1 What of Prior and Related Work?

To bring this section to suitable closure, we must address, at least briefly, an apparent incompatibility between the negative result we have obtained (i.e. the Uncomputability of Meaning theorem = UMT), and what some others have said regarding how much is demanded, computationally, for natural-language understanding (NLU). We must also point out that some approaches to NLU outside inferential semantics (as least avowedly so by what proponents of these approaches say; what might be the case formally is another matter) appear to be quite consistent with UMT. In this regard, we quickly mention three strands of related, prior research. In the first strand, NLU, while held to be "hard," is by definition computable because the level of difficulty is firmly within the Polynomial Hierarchy, and the background assumption appears to be some such proposition as a cognitive analogue to the Church-Turing Thesis. In the second strand of research, a very robust, cognitively realistic approach to NLU, we perceive at least apparent uncomputability in the general case. In the third strand, a connection is made between language acquisition formally modeled, and NLU. Here now is the (very brief) commentary on these three strands, respectively.

9.2 Descriptive Complexity

In general, descriptive complexity is the marriage of standard coverage of complexity theory with formal logic: formal logics (of the standard, extensional and bivalent sort) are used to describe the difficulty of computing functions, where—and this is crucial at present—difficulty consists in the size of demands for time and space to compute the functions in question. (Formal logics are also of course used in standard ways to specify both the Arithmetic ($\mathscr{L}_1$ used) and Analytic ($\mathscr{L}_2$ used) Hierarchies, which are dominated by problems that *aren't* computable.) From this perspective, according to which, by definition, all the functions in question are computable, in a recent paper that provides an insightful overview of the human mind and descriptive complexity, Pantsar [21] at least implicitly affirms the proposition that NLU is computable. What is the basis for the affirmation of such a proposition? As Pantsar nicely reports, a large part of that basis is Ristad's [24] claim that human-level natural-language computations are **NP**-complete—and hence by implication computable. As to not just a claim, but a *theorem* that forms part of the basis for the proposition as well, Pantsar cites Fagin's [10] theorem that a proper subset of full second-order logic = $\mathscr{L}_2$ suffices to describe **NP**. For the most part, all of this, and more, is orthogonal to the UMT result. Certainly none of this is at all a threat to, or even for that matter inconsistent with, UMT. The reason is simple: UMT is not rooted in anything like the *description* of problems in a given formal logic; rather, UMT is rooted in the treatment of meaning in terms of inference built from the proof- or argument-theory of formal logics. In this account of meaning, it's the inferential dimension of formal logics that is central, not a dimension relating to capture of functions by way of formulae, and—as is obvious from the rejection of Montagovian meaning issued above—not a dimension relating to model-based semantics.[11]

9.3 Computational Cognitive NLU

Here we take a recent volume, *Linguistics for the Age of AI* [19], as an exemplar. As far as we are aware, this work is the most robust use of cognition (computationally modeled) and declarative knowledge for NLU that takes the full challenge of real natural language (with e.g. all its ambiguity) seriously. The first author's contention is that NLU here is uncomputable, since the knowledge brought to bear in order to enable the understanding of some natural language is arbitrarily expansive in the general case, and the natural language to be understood is likewise unrestricted. If the knowledge here is captured by some set Γ in a formal logic, and if ultimately there is at least a set of reasons for why some natural language S is to have some meaning expressible in a formal logic as formulae Φ_S, then it's not hard to see why the overall problem might be uncomputable, along the lines of our own framework for $\mathscr{HIS}$, and for UMT.

[11] Some readers may naturally ask whether some direct treatments of natural language by symbolizing/representing that language in formal logics have been shown to lead to uncomputability. While this is far beyond our scope here, we find it noteworthy that in his treatment of quantification in human natural language, Szymanik (see e.g. [27]) explicitly treats the understanding of this language to be a computable affair. In fact, he appears to affirm the proposition that human/human-level cognition overall is Turing-computable, something that the first author has long rejected, and defended repeatedly in print (e.g. see [2, 6]).

9.4 Computational Learning Theory

Computational learning theory (CLT), a firmly recursion-theoretic approach to machine learning (and hence radically different than today's data-driven "ML"), might seem to some readers to be quite relevant to UMT. In CLT, the focus is on *language acquisition*; that is, the challenge is for some agent to acquire, through time, command over a language. The language in question is identified with a Type-0 grammar; this means that what is to be learned can simply be identified with a Turing machine. The *locus classicus* of CLT is [17]. This work is a litany of limitative theorems, including those that say that learning a Turing machine, based on perceiving only small, finite snippets of information regarding the machine in question, is not a computable challenge. But why might CLT be thought relevant to UMT? The reason is simply that someone might view NLU in a broader way than we do, to specifically include, first, the understanding of what grammar/Turing machine is in play, and then, following on that, something more specific, and specifically connected to how we define NLU. We in general certainly see the reasonableness of having such an extended conception of NLU, and we appreciate that CLT is exceptionally difficult, but our definition of NLU in inferential terms is such as to only make CLT vaguely relevant to the theorem UMT. In fact, when in CLT a given agent is given information about the target to be learned, no inference whatsoever is in play, and indeed the agent is said to be successfully learning the target if and only if hypotheses about what that target is are wrong only finitely many times in the limit. No proof or argument is in the picture at all.

10 Future Work; Further Objections

At this point $\mathscr{HIS}$ is admittedly really only a proto-theory of meaning, and only the basis for *future* NLU and NLG. There is long and hard work ahead. But fortunately, prior work by Bringsjord and colleagues in robot ethics (of the sort encapsulated in Sect. 3) is an ideal foundation upon which to develop and refine, mathematically, $\mathscr{HIS}$, and NLP algorithms and technology built upon this approach. The reason why is obvious: the approach to robot ethics in question is steadfastly and thoroughly proof- and argument-centric; and since on $\mathscr{HIS}$ the meaning of natural-language sentences are captured by proofs and arguments that contain then, and are expressed in the form of (inevitably vast) hypergraphs that express these interacting proofs and arguments, the fit is a most promising one.

A final word regarding additional objections to $\mathscr{HIS}$ and to building NLP upon automated reasoning and proof-/argument-checking with the relevant hypergraphs: We are under no such illusion as that additional objections will not be pressed against the formal semantics advanced above. Inevitably, many will object that the meaning of a natural-language sentence s is captured by "static," non-inferential content expressed in some formal logic (or, equivalently, in some knowledge-representation format, say frame-based representations). Such skeptics would do well to consider the longstanding fact that in formal logic, the capture of a given mathematical assertion a, which is part natural language and part formal language, has for many decades been known to be achievable not merely by producing a formula $\phi(a)$ in some formal language that *expresses* a, but by a formal theory Φ which is such that $\phi(a)$ is provable from Φ. A very nice presentation of the express-vs-capture distinction when applied to statements in mathematics is provided in Chap. 4 of [26].

Though as we readily admit there is much work to be done, we recommend that household robots (and *a fortiori* robots that as a matter of course frequently find themselves in morally charged situations, e.g. military robots) be engineered to process and interactively discuss norms with humans on the basis of the computational logics and corresponding procedures that underlie $\mathscr{HIS}$.

Acknowledgements We are indebted to multiple commentators on the presentation at ICRES 2020 of a proper subset of work herein described, and on the older, smaller paper that the present one subsumes, augments, and refines. Without long-term support of research and development in the Rensselaer AI & Reasoning Laboratory that centers around inference, the formal semantics adumbrated in this paper would not exist, and we are very grateful for this support from AFOSR (award # FA-9550-17-1-0191) and ONR in the States. We are indebted to two anonymous referees for cogent comments on an earlier kernel of the present paper, and to both Bertram Malle and Paul Bello for *substantive* comments, advice, and critique. Finally, the lead author thanks Bill Rapaport, arguably the world is leading light regarding Buffalo sentences (and perhaps Buffaloian cuisine as well), for many enlightening conversations.

References

1. Bringsjord S (1985) Are there set-theoretic worlds? Analysis 45(1):64
2. Bringsjord S, Arkoudas K (2004) The modal argument for hypercomputing minds. Theor Comput Sci 317:167–190
3. Bringsjord S, Govindarajulu N (2020) The theory of cognitive consciousness, and Λ (Lambda). J Artif Intell Conscious 7(1):155–181. http://kryten.mm.rpi.edu/sb_nsg_lambda_jaic_april_6_2020_3_42_pm_NY.pdf (The URL here goes to a preprint of the paper)
4. Bringsjord S, Govindarajulu NS (2012) Given the web, what is intelligence, really? Metaphilosophy 43(4):361–532. http://kryten.mm.rpi.edu/SBNSGRealIntelligence040912.pdf (This URL is to a preprint of the paper)
5. Bringsjord S, Taylor J (2012) The divine-command approach to robot ethics. In: Lin P, Bekey G, Abney K (eds) Robot ethics: the ethical and social implications of robotics, MIT Press, Cambridge, MA, pp 85–108. http://kryten.mm.rpi.edu/Divine-Command_Roboethics_Bringsjord_Taylor.pdf
6. Bringsjord S, Kellett O, Shilliday A, Taylor J, van Heuveln B, Yang Y, Baumes J, Ross K (2006) A new Gödelian argument for hypercomputing minds based on the busy beaver problem. Appl Math Comput 176:516–530
7. Bringsjord S, Taylor J, Shilliday A, Clark M, Arkoudas K (2008) Slate: an argument-centered intelligent assistant to human reasoners. In: Grasso F, Green N, Kibble R, Reed C (eds) Proceedings of the 8th international workshop on computational models of natural argument (CMNA 8), University of Patras, Patras, Greece, pp 1–10. http://kryten.mm.rpi.edu/Bringsjord_etal_Slate_cmna_crc_061708.pdf

8. Dummett M (1991) The logical basis of metaphysics. Duckworth, London, UK
9. Ebbinghaus HD, Flum J, Thomas W (1994) Mathematical logic, 2nd edn. Springer, New York, NY
10. Fagin R (1974) Generalized first-order spectra and polynomial-time recognizable sets. In: R Karp (ed) SIAM-AMS Proceedings complexity of computation, vol 7, pp 43–73
11. Francez N (2015) Proof-theoretic semantics. College Publications, London, UK
12. Gentzen G (1935) Investigations into logical deduction. In: Szabo ME (ed) The collected papers of Gerhard Gentzen, North-Holland, Amsterdam, The Netherlands, pp 68–131. (This is an English version of the well-known 1935 German version)
13. Govindarajulu N, Bringsjord S (2017) On automating the doctrine of double effect. In: Sierra C (ed) Proceedings of the Twenty-Sixth international joint conference on artificial intelligence (IJCAI-17), international joint conferences on artificial intelligence, pp 4722–4730. https://doi.org/10.24963/ijcai.2017/658
14. Govindarajulu N, Bringsjord S, Sen A, Paquin J, O'Neill K (2018) Ethical operating systems. In: De Mol L, Primiero G (eds) Reflections on programming systems, philosophical studies, vol 133. Springer, pp 235–260. http://kryten.mm.rpi.edu/EthicalOperatingSystems_preprint.pdf
15. Govindarajulu NS, Bringsjord S (2015) Ethical regulation of robots must be embedded in their operating systems. In: Trappl R (ed) A construction manual for robots' ethical systems: requirements, methods, implementations, Springer, Basel, Switzerland, pp 85–100. http://kryten.mm.rpi.edu/NSG_SB_Ethical_Reg_at_OS_Level_offprint.pdf
16. Govindarajulu NS (2016) ShadowProver. https://naveensundarg.github.io/prover/
17. Jain S, Osherson D, Royer J, Sharma A (1999) Systems that learn: an introduction to learning theory, 2nd edn. MIT Press, Cambridge, MA
18. Kleene S (1967) Mathematical logic. Wiley, New York, NY, I recommend a 2002 Dover unabridged republication of the original 1967 book from Wiley, if you cannot obtain the original book
19. McShane M, Nirenburg S (2021) Linguistics for the age of AI. MIT Press, Cambridge, MA
20. Montague R (1974) Formal philosophy: selected papers of Richard Montague. Yale University Press, New Haven, CT, Note that this book is the seminal work of author Montague, but is edited by Thomason
21. Pantsar M (2021) Descriptive complexity, computational tractability, and the logical and cognitive foundations of mathematics. Minds Mach 31:75–98. https://doi.org/10.1007/s11023-020-09545-4
22. Prawitz D (1972) The philosophical position of proof theory. In: Olson RE, Paul AM (eds) Contemporary philosophy in Scandinavia. Johns Hopkins Press, Baltimore, MD, pp 123–134
23. Quinn P (1978) Divine commands and moral requirements. Oxford University Press, Oxford, UK
24. Ristad E (1993) The language complexity game. MIT Press, Cambridge, MA
25. Schroeder-Heister P (2012/2018) Proof-theoretic semantics. In: Zalta E (ed) The stanford encyclopedia of philosophy. https://plato.stanford.edu/entries/proof-theoretic-semantics
26. Smith P (2013) An introduction to Gödel's theorems. Cambridge University Press, Cambridge, UK, This is the second edition of the book
27. Szymanik J (2009) Quantifiers in TIME and SPACE: computational complexity of generalized quantifiers in natural language. University of Amsterdam, Amsterdam, The Netherlands, This is a dissertation published by Institute for logic, language and computation, dissertation series DS-2009-01

Reinventing Kantian Autonomy for Artificial Agents: Implications for Self-driving Cars

Endre Erik Kadar and Zsolt Palatinus

Abstract Robot Ethics is a new research area within Applied Ethics. One of the central problems in Robot Ethics is the possibility of regarding artificial agents as moral agents. The present paper is considering this challenge and proposes a framework inspired by history, specifically we argue for the need to reinvent autonomy for artificial moral agency based on Kantian ethics. In addition to autonomy, other key notions of Kant's system such as rationality, laws of freedom are critically discussed. The feasibility of proposal includes examples from existing practices such as human use of animals as autonomous agents and reinterpretation of moral behaviour (rationality, autonomy, freedom, etc.) in autonomous vehicles.

1 Introduction

Robot Ethics is a new research area in Applied Ethics. It has emerged as a result of the increasing complexity of Artificial Agents including mobile robots. There seems to be a consensus that a new class of robots that are able to perform increasingly complex tasks independently of human control are different than previous robots to the extent that their behaviour should be ethical. These robots are often called autonomous systems (artificial agents) that are able to make decisions in complex tasks and their behaviour could also satisfy moral principles of behaviour control. However, when researchers are trying to identify ethical principles of robot behaviour for specific robots it is much harder to find consensus beyond the abstract laws of Asimov that are defined in relation to humans. In short, this relationship is based on a meta principle, which can be put simply: Robots should behave as if they are enslaved to humans

E. E. Kadar (✉)
Department of Psychology, University of Portsmouth, King Henry Bldg., King Henry I Street, Portsmouth PO1 2DY, UK
e-mail: endre.kadar@port.ac.uk; kadar_e@yahoo.co.uk

Z. Palatinus
Department of Cognition and Neuropsychology, Institute of Psychology, University of Szeged, Egyetem u. 2, Szeged 6722, Hungary
e-mail: palatinus.zsolt@psy.u-szeged.hu

© The Author(s), under exclusive license to Springer Nature Switzerland AG 2022
M. I. A. Ferreira and M. O. Tokhi (eds.), *Towards Trustworthy Artificial Intelligent Systems*, Intelligent Systems, Control and Automation: Science and Engineering 102,
https://doi.org/10.1007/978-3-031-09823-9_12

and robots should never endanger humans. This meta principle sounds plausible but designing robots that are compliant and truly enslaved to humans has its danger. Asimov's laws, in principle, could also be accepted as laws of human behaviour but they are clearly insufficient for guiding human behaviour in everyday life which is often full of tensions and conflicts arising from different motivations, interests, and expectations of individuals. The present paper offers a broader perspective on the state of the art in Robot Ethics by revealing that the current situation with controversies in seeking a foundation for ethical behaviour for robots is similar to the eighteenth century debates on ethical principles and theories in finding a broad and consistent theory for human moral behaviour. In the eighteenth century, autonomy became the central notion as Kant elucidated nicely in his effort in seeking a foundation of ethics for human moral agents. The present paper argues that a Kantian approach (i.e., agents with autonomy) is needed to provide a foundation for moral behaviour in artificial agents. In other words, researcher of Robot Ethics need to reinvent the Kantian autonomy for robots. The arguments in support of this approach are based on historical analogy, existing praxis of dealing with tamed and domesticated animal behaviour. Finally, we demonstrate the benefits of this approach in self-driving cars.

1.1 Robot Ethics as Applied Ethics

Robot Ethics is a new research area within Applied Ethics but this emerging area of research is different to other areas of Applied Ethics. Why? The notion of ethics is used in three different but related ways: (1) general patterns of behaviour, the way individuals behave in their natural and social environment, (2) A set of rules (moral codes) of the conduct of behaviour in a society, and (3) investigation of the way of life, the rules, norms, and principles (codes) of behavior [1]. Applied Ethics is the branch of Ethics dealing with the analysis of specific, controversial moral issues and as such Applied Ethics is encompassing all above three meanings of Ethics. Controversial issues in behaviour imply the lack of consensus that leads to conflict on the interpretation of what is the right and wrong behaviour under specific conditions. In other words, inconsistencies in rules and moral codes are present in problems of Applied Ethics. These are usually embedded into everyday life in which most behaviour patterns are considered natural without moral concerns even though they also tacitly based on fundamental moral codes of conduct.

By looking at the principles that are most commonly used in Applied Ethics summarized by Fieser [6]:

- *Personal and social benefits*: acknowledge the extent to which an action produces beneficial consequences for a person or a society.
- *Principles of benevolence and paternalism*: help those who are in need of help especially if they cannot help themselves.
- *Principles of no-harm, honesty*: do not harm and deceive others.
- *Principle of lawfulness*: do not violate the law.

- *Principles of justice and rights of persons*: acknowledgment of the right to due process, fair compensation for harm done, and fair distribution of benefits, rights to life, information, privacy, safety and free expression of views.
- *Principle of autonomy*: acknowledge a person's freedom over his/her actions or physical body.

These principles represent a range of traditional normative principles including both consequentialist and duty-based ones. The personal and social benefits are consequentialist since they appeal to the consequences of an action as it affects the individual or society. The remaining principles are duty-based. The principles of benevolence, paternalism, harm, honesty, and lawfulness are based on duties we have toward others. The principles of justice, and the various rights are based on moral rights and they require human person as agent. Importantly, the principle of autonomy is usually based on the assumption that the agent has personality which implies human agent. In other words, the last three principles are not readily applicable to artificial agents. The present paper is focusing on the principle of autonomy in order to find a plausible ground for defining artificial moral agency with autonomy rather than taking the position that artificial moral agency is beyond science and the notion of moral agency can only be applied to humans. Our argument is based on a critical discussion of basic concepts such as agency and moral agency. To bridge the obvious gap between humans and artificial agents, animal behaviour patterns and the possibility of animals as agents with moral behaviour are presented.

1.2 Agency

The notion of *agency* is associated with the idea of being a source of an action or an initial step of a chain of events that cannot be reduced to other earlier events. To put it simply, being an agent in the broadest sense is to be a source of irreducible causal factor in the flow/chain of events. Aristotle's four types of causation could be recalled to provide insights into what type of causes we can think of (e.g., a chemical agent such as poison of a drug for treatment is a material cause, a force that pushes and move something is an efficient cause, etc.). Final cause plays a special role in defining agency in the narrower sense. It is the basis of the most common definition of agency. Accordingly, only beings capable of goal setting and goal directed action called agents. In other words, only *beings with intentional states* (i.e., mental states that are *about* something else, like a desire for a future state), *are regarded agents*. Clearly, humans are the prime examples for agency but there are many living creatures such as primates, elephants, and dogs that are also capable of performing acts or initiating a chain of events because they are capable of having intentional states and act in order to achieve an intended future state. In other words, they are able to perform goal-directed action.

1.3 Moral Agency

Obviously, the class of moral agents is narrower than the class of agents. The notion of agency is different from *moral* agency that is associated with the idea of being accountable for one's behavior. Being *accountable* (or morally responsible) for behaviour implies that an agent's behavior is governed by moral standards and hence that one has moral duties or moral obligations. In other words, one's behaviour should be guided by and hence evaluated under those standards.

This notion is usually regarded as widely accepted and non-controversial. For instance, according to *Routledge Encyclopedia of Encyclopedia*, "[m]oral agents are those agents expected to meet the demands of morality" [8]. Similarly, *Stanford Encyclopedia of Philosophy* states that "a moral agent [is] one who qualifies generally as an agent open to responsibility ascriptions" [5]. Also, the *Internet Encyclopedia of Philosophy,* defines moral agents that "can be held accountable for their actions—justly praised or blamed, deservedly punished or rewarded" [13].

Thus, the idea of moral agency implies that it is appropriate to hold moral agents accountable for their behavior. Accountability implies that there are accepted social conventions that includes norms/standards of behaviour and they are rational/transparent and they are enforced (i.e., socially justified to reward right behaviour and sanction wrong behaviour). One can immediately recognize that moral standards are not absolute. Different ages and different cultures used different moral standards to hold moral agents accountable for their behavior. This recognition may lead to moral relativism and raises the possibility of arbitrariness of conventions different societies developed as their system of moral standard. There are examples for various religions with standard based on some holy insights (religious illumination) by a prominent person, a leader. These examples seem to confirm arbitrariness constrained by divine insights. On the other hand, living conditions of a society are given and the everyday activity of the society has to be adequate for the conditions rather than arbitrary. For instance, harsh climate and difficulties in having sufficient resources such as food are often associated with rigid and uncompromising moral standard to avoid wasteful and unjust use of resources. Having said that, it is obvious that there is a certain degree of freedom (arbitrariness) in social norms, the standard of behaviour. This seems analogous to biological adaptation to new conditions with some variance that allows flexibility to adjust to smaller range of changing conditions.

Another important and usually unquestioned assumption in the creation and use of moral standards is that the agents are free and they can choose to act in accordance with the standard or deviate from the expected behaviour patterns in everyday life. Free agents also possess the ability to act rationally but there might be conflicts arising between the rationality of the system of (social) norms and the subjective rationality of an agent (a member of a society). *Rationality* of an agent often implies *consciousness* because unconscious actions and decisions to select actions are automatic and as such the agent cannot be responsible for unconscious actions. For instance, in sleepwalking or under hypnotic control an agent is not in a normal waking state and not in full control, the agent is not fully responsible for his/her actions.

1.4 Can Animals Be Moral Agents?

Based on the above discussed criteria of moral agency, non-human animals such as dogs, cats and elephants are obviously agents but they cannot be considered as moral agents [9]. They are not rational and do not have human-like conscious control of behaviour. Despite these obvious differences between humans and non-human animals, many animals are domesticated to such a degree that their behaviour patterns are adjusted to their coexistence with humans. In some cases, human trust in animals can reach very high level. For instance, using guide dogs humans put their life at risk by transferring behaviour control to the animal partner (guide dogs are working in strong synergy with their blind owners in navigating in complex environments). Also, humans are using elephants that are far stronger than humans and could crush a human easily, yet, elephants are carrying humans on their back and they often also help carry heavy loads. In general, humans train various animals to become agents whose behaviour patterns fit into the world of humans. These animals do not have to have rational knowledge of human behavioural norms. In everyday life, it is not considered important to have awareness of norms. It is sufficient to ensure that behaviour patterns are adjusted to fit into the dynamics of the life-world of humans. Thus, these examples of coexistence with domesticated animals suggest an alternative approach to Animal Ethics which could be used in Robot Ethics. It would be sufficient to have artificial agents that are similarly reliable to trained animals (e.g., robots to guide a blind person, or carrying heavy loads on their back including humans while navigating safely in the environment). It seems plausible to give up on the stringent requirements of having human-like mental capacity for being a moral agent. Instead, there is a need for an alternative framework for autonomous robots that are able to behave according to similar standards to some (domesticated/pet) animals who behave safely and even morally sound in a shared life-world with humans. To put differently, robots are required to be trained to reach the level of reliability of some domesticated animals so they can be allowed to move into human habitat without posing danger to humans or raising ethical concerns. Based on these examples, moral artificial agents, ethical robots could be defined based on the criteria used for animals regardless of their different moral status than humans have according to the common interpretation of moral agency.

Animals are obviously agents and many of them (dogs, cats, elephants, dolphins, etc.) have intentional disposition (state) that is vital in goal directed behaviour. Typically, animals are able to adapt and they adjust/change their behaviour patterns if conditions require to do so. These adaptation processes are taking place at different time scales and some of these processes are for the purpose of adjusting behaviours to humans (e.g., dogs evolved from wolves to become companion of humans, wild horses and elephants are tamed to live with and work for humans). Many wild animals who normally tend to stay away from humans develop a new approach and become city dwellers and stop worrying about keeping safe distance from humans (e.g., squirrels, pigeons are looking for dropped pieces of food or even begging for food from eating humans). These examples of changing behaviour patterns in animals are clear evidence for animal's autonomy because animals create new behaviour patterns based on their own terms. This is autonomy in its original sense—self-defined rules/laws of behaviour. In sum, animals have the freedom to choose behaviour patterns and they have autonomy to define patterns of behaviour by either following patterns of previous generations of the species or changing these patterns or creating radically new ones based on their needs and the circumstances. These behaviour patterns could be considered moral behaviour patterns without explicit moral standards. This view on animal behavior can be regarded as a non-reflective ethics without codification of norms, which are also typical in early tribal and other primitive human cultures. Also, this approach can be used to explain early developmental phases of ethical behavior in humans [4] or moral behavior of animals [3].

1.5 Autonomous Robots

There are three different meanings of ethics as discussed above [1]. The primary meaning is "general patterns of behaviour, the way individuals behave in their natural and social environment". Simply, at the fundamental level ethics is not complicated by reflective conscious processes, instead, ethical conduct is simply behaviour patterns that suits for coexistence with other humans. Arguably, the seemingly "primitive" structure of behavior patterns are well-suited for social life without thinking about canonizing behavioral standards to ensure coherence of principles and norms. With increasingly complex social structures that are associated with increasingly complex behaviour patterns the reflection on these patterns lead to the need for seeking standards, principles or norms that would inform and guide old as well as new behaviour patterns. Early human societies are not based on complex codified system of norms. Behaviour patterns have emerged in accordance with the demand of surviving in natural habitat of a social group. These patterns included coordinated social activities.

Over time, humans started using various animals including dogs, horses, etc. and the behaviour patterns of these animals are based on training without teaching "ethical codes" for animals. In sum, ethics is not theoretical at the basic level of simple behaviour patterns. By way of analogy, at the early stage of developing autonomous robots that are seamlessly fit into behaviour patterns of human cultures, robot behaviours can also be based on focusing on behaviour patterns by training just like infants or pets are trained and they are learning suitable behaviour patterns to fit in their environment. Training can be passive as well as active but mostly it is a mixture of both. Active training implies complex learning capability and that should be the focus of contemporary Robot Ethics rather than focusing on implementing an abstract codified system of norms/principles of behavior control.

This process of learning/training can be regarded as one possible interpretation of the Kantian ethics. The key idea of this system is autonomy, the freedom of the agent, the freedom of defining its laws of motion. This true sense of autonomy is not arbitrary because it is constrained by the Kantian maxim. According to the maxim, these individually defined laws of behavior should universal rather than unique, subjective and valid for the individual only. To be able to apply this constraining factor to defining laws of behavior requires rationality. These requirements of the Kantian approach are often regarded as something that is practically impossible to satisfy by humans. These are indeed stringent conditions according to the traditional meanings of laws and rationality. However, if we give up on using traditional meaning of rationality and accept bounded rationality [12] we can provide a blue-print that could be used for defining a key aspect of autonomy of artificial agents. More specifically, rationality can be interpreted as adaptation instead of an abstract mental assessment of all possible behavioral consequences [7]. In terms application of our proposed Kantian framework to robot behaviour, the complexity of the laws of behavior patterns can be further reduced by the limitations/constraints of possible situations a robot may encounter. In sum, this is a surprisingly simple pragmatic approach to avoid all the abstract theoretical debates that often arise in search of principles/norms for robot behavior. The feasibility of this proposal could be demonstrated in deriving movement control strategies of self-driving vehicles.

2 Self-driving Vehicles

The proposed Kantian autonomy-based ethics can be applied to self-driving vehicles. Self-driving cars are different than robots because humans are on board of these vehicles. Thus, the movement control has to be adjusted to the traffic conditions as well as to the passengers to ensure that they fell safe and comfortable with the movement of the vehicle. There is an apparent paradox in our proposed approach. On the one hand, it seems abstract and highly theoretical and theories of driving do not suggest simple solutions to bridge the gap between human driving performance and movement control of self-driving cars [10]. On the other hand, adaptation/learning to imitate/simulate drivers behavior could be primarily an empirical and technical rather than a theoretical problem. How could this work? First of all, the Kantian approach to moral behaviour is difficult because its universality requirement and consistency [11]. Specifically, self-driving vehicles are usually envisioned as universal artificial driving systems that are able to navigate traffic conditions human drivers encounter. This universality is appealing but at the same time it seems overly ambitious. Arguably, it is unnecessarily ambitious because the environment drivers and the self-driving vehicles are using regularly are mostly limited. Also, different people use different roads and encounter different traffic conditions during their typical daily driving routine. It is well known that driving styles could be quite different for different individuals but it is also easy to overlook important differences between drivers because of different personality traits. Individual differences could also be influenced by many other factors (e.g., visual acuity, experience in driving, fatigue, driving conditions such as visibility, road conditions, etc.).

If an autonomous driving system is used by an individual and the control of driving is uniformly defined based on the manufacturer's design, the user of this system need to get used to the movement control system. Some individuals may find it difficult to get used to this "artificial" control and may find it hard to notice if there is a possible malfunctioning in the auto-pilot system. Based on these concerns, it seems plausible to adjust the "laws of behavior" of the artificial driving system to be similar to the driver's style of control to ensure the driver would feel comfortable with the "robot driver" and feel familiarity with the movement control under various conditions. This requirement seems challenging but, in fact, this can help simplify the problem by using driving patterns of a person to derive abstract rules of driving for the robot driver under different classes of conditions (straight roads in inner-city, highway driving with multiple lanes, rural meandering roads, residential areas, various visibility and road conditions, etc.). Of course, the learning of an individual's driving style implies that the individual has considerable driving experience and definitely not a novice driver. Another simplifying factor in the process of defining driving behavior for a self-driving vehicle is the use of GPS system and road maps. These aids could be used to split the overall trajectory of a journey into segments (straight road, approaching a bend, approaching an intersection, exiting highway, etc.) for which laws of driving (safe range of speed, acceleration, deceleration, car following distance, etc.) could be set differently. All of these are basically adaptation processes in terms of learning the style of a specific driver that includes deriving laws of behavior control under various traffic conditions of a specific human driver (a user of the autopilot system). In the Kantian language, rationality of defining laws of behavior is adaptation of a specific individual's driving style based on smart artificial learning algorithms that could avoid using conflicting theoretical perspectives on driving [10] and the puzzle of application of universal ethical norms [2].

3 Conclusions

The present paper discussed Robot Ethics as a new research area within Applied Ethics. What makes Robot Ethics different than other areas of Applied Ethics is the special status of agency. Specifically, the increasingly important problem of artificial moral agency was critically reviewed and presented by taking a historical perspective. The present paper proposes a framework inspired by the similarity of situation with regard to debates on principles of Ethics during the eighteenth century and present days. The difference is that in the eighteenth century varieties of theories emerged with increased uncertainty on finding a widely accepted definition of humans as moral agents while today the focus is on finding moral standards for artificial moral agency. Toward the end of eighteenth century Kant came up with a radical proposal by defining moral agency on the basis of individual autonomy of his/her behavioural conduct. The present paper argues for the need for reinventing autonomy and using it for artificial moral agency based on Kant's approach to ethics. In the process of reinventing autonomy, other key notions of Kant's system such as rationality, laws of freedom have been critically discussed. The feasibility of this proposal demonstrated by examples from existing practices such as human use of animals as autonomous agents with a kind of "moral behaviour" and reinterpretation of rationality in moral behaviour for human agents and autonomous vehicles. In this proposal, adaptation processes are playing the role of rationality in deriving movement control strategies for natural and artificial moral agents.

References

1. Abelson R, Nielsen K (1967) History of ethics. In: Edwards P (ed) Encyclopedia of philosophy. Macmillan, New York
2. Allen C, Varner G, Zinser J (2000) Prolegomena to any future artificial moral agent. J Exp Theor Artif Intell 12(3):251–261
3. Bekoff M (2004) Wild justice and fair play: cooperation, forgiveness, and morality in animals. Biol Philos 19(4):489–520
4. Dahl A (2014) Definitions and developmental processes in research on infant morality. Hum Dev 57:241–249
5. Eshelman A (2010) Moral responsibility. In: Zalta EN (ed) Stanford encyclopedia of philosophy. http://plato.stanford.edu/entries/moral-responsibility/
6. Fieser J (2020) Ethics. The Internet encyclopedia of philosophy. ISSN 2161-0002. https://www.iep.utm.edu/
7. Fleurbaey M (1988) Rational behaviour and adaptation. In: Munier BR (ed) Risk, decision and rationality. D. Reidel Publishing Company, pp 459–481
8. Haksar V (1998) in *Routhledge Encyclopedia of Philosophy*. Available at https://www.rep.routledge.com/
9. Himma KE (2009) Artificial agency, consciousness and the criteria for moral agency: what properties must an artificial agent have in order to be a moral agent? Ethics Inf Technol 11(1):19–29
10. Kadar EE (2019) Mind the gap: a theory is needed to bridge the gap between the human skills and self-driving cars. In: Ferreira MIA, Sequeira JS, Virk GS, Tokhi MO, Kadar EE (eds) Robotics and well-being. Springer, pp 55–65
11. Powers TM (2006) Prospects for a Kantian machine. IEEE Intell Syst 21(4):46–51
12. Simon HA (1988) Bounded rationality and organizational learning. Technical report AIP-107
13. Williams G (2020) Responsibility. The Internet encyclopedia of philosophy. ISSN 2161-0002. https://www.iep.utm.edu/

Realizing the Potential of AI in Africa: It All Turns on Trust

Charity Delmus Alupo, Daniel Omeiza, and David Vernon

Abstract Most nations have recognized the disruptive influence of artificial intelligence (AI) on all aspects of their economies, from manufacturing, to services, to governance, and the potential benefits that embracing AI technologies can bring. It is no different in developing countries and it is certainly the case that the countries of Africa have embraced AI, data science, machine learning, and robotics. However, the transition from recognition of potential to realization of benefits is not a straightforward matter. In this essay, we argue that this transition depends on turning technological invention into innovation, that technological innovation cannot happen without adoption, and that adoption depends on socio-cultural factors, in general, and on trust, in particular. We draw out the implications for AI in developing countries in Africa, arguing that, for Africa to realize the potential of AI in solving economic and social problems, the advancement and deployment of AI must be driven and executed by the peoples of Africa: if it is not, there will be little trust, less adoption, and minimal benefits.

1 Introduction

Humankind has always used tools to augment and amplify its capability for physical work. The computer age extended this to mental work but principally as a tool for greatly increasing the speed of processing. However, the developments in artificial intelligence (AI) over the past sixty five years, and in particular in deep learning

C. D. Alupo
School of Business, University College Dublin & SFI Centre for Research Training in Machine Learning, Dublin, Ireland
e-mail: charity.alupo@ucdconnect.ie

D. Omeiza
Department of Computer Science, University of Oxford, Oxford, UK
e-mail: daniel.omeiza@cs.ox.ac.uk

D. Vernon (✉)
Carnegie Mellon University Africa, Kigali, Rwanda
e-mail: vernon@cmu.edu

© The Author(s), under exclusive license to Springer Nature Switzerland AG 2022
M. I. A. Ferreira and M. O. Tokhi (eds.), *Towards Trustworthy Artificial Intelligent Systems*, Intelligent Systems, Control and Automation: Science and Engineering 102,
https://doi.org/10.1007/978-3-031-09823-9_13

approaches in the past ten, have ushered in what John Kelly at IBM refers to as the cognitive era [50], superseding the tabulating era and the ongoing programming era. He reminds us of J. C. R. Licklider's prediction in 1960 of a symbiotic partnership between man and computer that will "perform intellectual operations much more effectively than man alone can perform them" [56]. Today, that symbiotic partnership is realized through AI, a tool that both amplifies and extends human cognitive abilities.

While there is much debate about the ultimate destiny of that partnership and whether or not AI will prevail over humans after the technological singularity when the autonomous capabilities of AI exceed those of humans [78], there is far more discussion on a more practical level about how to harness AI for economic advantage and social development.

From an economic standpoint, AI forms the foundation of the so-called Fourth Industrial Revolution, Industry 4.0 [77]. Concerned about missing out on this revolution, countries and trading blocs around the world have prepared or are in the process of preparing AI strategies. Unsurprisingly, most of this activity occurs in developed economies, in the West, e.g. [3], Middle East, e.g. [7], and the Far East, e.g. [4],[1] where governments and organizations in the public and private sector seek to ensure that they have both a continued competitive advantage in global trade while maintaining or improving the standard of living of their citizens in economic and societal terms. The scope of these strategies is extensive. It embraces the research and development necessary to advance even further the capability of AI in disciplines, ranging from data science, to machine learning, to cognitive robotics, and much else in between. It also embraces strategies for promoting innovation—entailing the widespread adoption of scientific breakthroughs and related engineering inventions in AI—and ensuring that this is done in a trustworthy, ethical manner [28].

Amidst all this vigorous effort to investigate and exploit AI in developed countries, the relevance to developing countries is often neglected, except for a few cases where some agencies seek to assist developing countries, e.g. [23, 29], or initiatives taken in the developing countries themselves, e.g. [5]. Consequently the special role that AI can play in supporting developing countries doesn't always receive the attention it merits [88]. This essay attempts to shift the spotlight and highlight the ways in which AI is relevant for developing countries, focusing specifically on Africa.[2] In this essay, we refer mainly to the countries in Sub-Saharan Africa, that is, East Africa, West Africa, and Southern Africa. In doing so, we highlight the factors that influence the success or failure of efforts to leverage the benefits that AI can provide. As we will see, it all turns on trust and the crucial difference between invention and innovation.

[1] For an overview of national AI policies and strategies globally, see [69].

[2] Africa is a continent comprising fifty-four countries and many different cultures: it is not a homogeneous society and this needs to be borne in mind when speaking of "Africa". Indeed, the words of Horst Köhler, former President of Germany, in his speech on the impossibility of speaking of Africa sounds a cautionary note: "I would like, if I may, to clear up one misunderstanding right away: Horst Köhler is not an Africa expert ... the reality on the ground in Africa is so much more complex than written accounts suggest ... the more I learned about Africa, the more I realised how much there was still to learn" [51].

2 The Benefits of AI Depend on Adoption and Trust

Rose [76] distinguishes between creativity, invention, and innovation. Creativity can lead to the invention of a novel idea or artefact but innovation carries the creativity and inventions into wider use: the diffusion of that invention and its widespread adoption, leading to substantial social change in the practices of a community of people. He frames this succinctly in a formula: "innovation = invention + exploitation + diffusion", where the invention is commercially developed and exploited, and, significantly, adopted in a wider community of users.

Successful innovation also depends on infrastructure. Rose notes that infrastructure is the unnoticed precondition for technology innovation [76]. There are two forms of infrastructure, the physical and the social. The physical infrastructure might include the availability of electrical power, communications networks, or internet connectivity, something that is taken for granted in developed countries but which cannot always be assumed in developing countries. Of equal importance is the social infrastructure which includes the social conventions that govern people's behaviour and the practices they find acceptable and unacceptable. Social infrastructure heavily impacts on whether or not an invention is adopted and becomes an innovation that can yield benefits for the local community. Crucially, social infrastructure includes trust and people's sense of what is trustworthy.

Hofman et al. [38] define trust as the expectation that a service will be provided or a commitment will be fulfilled, emphasizing the importance of *expectation* in their definition. Expectations, as they point out, depend on many things but, irrespective of what these expectations are, they are certainly grounded in the socio-cultural experience of those whose trust is required.

The importance of the cultural context in the evolution of trust is also emphasized by Lee and See [55], who define trust as "the attitude that an agent will help achieve an individuals goals in a situation characterized by uncertainty and vulnerability". They point out that there are many factors at play in the development of trust, including individual, organizational, and cultural context. They define culture as "a set of social norms and expectations that reflect shared educational and life experiences associated with national differences or distinct cohorts of workers". An awareness of these social norms and expectations, and the socio-cultural background from which they arise, is crucial to the development of trust in any new technology, including AI-based products and services, and by extension to their diffusion and adoption. For example, language as an important element of culture presents itself in multiple forms in Africa. There are about 1,500–2,000 officially recognised languages spoken across the 3,000 ethnic groups in Africa [70]. One needs to be alert to the biases that might surface if developers from a primary region, language, or tribe created a loan allotment algorithm for the remainder of the continent. This is worthy of concern because such scenarios have played out badly in other continents. In the USA, AI algorithms have been used in predictive policing, labeling residents based on location, social economic status, and name, resulting in cases where African Americans have been profiled, even though no crime was ever committed, no criminal record existed, and there was no recent contact with law enforcement [57].

Culture can be characterised in many ways. Hofstede identifies six dimensions in which an understanding of cultural issues should be addressed [39–41]. Others highlight the different ways that cultures perceive time and space, noting that concepts of time in the West and in Africa differ significantly [11]. Without wanting to fall into the trap of generalising across a multitude of cultures and ignoring ethnographic diversity, one can say that time in Africa has traditionally been tied to events, which may be regular or irregular, in contradistinction to the view in the West of time as continually moving from past, to present, to future. These factors have a bearing on how technology, generally, and information technology, powered by AI, in particular, can support an individual or a local community in Africa and whether or not that support, no matter how well intended, will be accepted, trusted, and adopted. Lack of trust can severely and negatively impact the adoption of these services and products, fatally undermining the achievement of the anticipated benefits: "Changes in the factors that affect users expectations will also impact users trust levels" [38]. Furthermore, AI brings its own special factors, e.g. explainability, transparency, lack of bias, all of which have their own influence on whether or not products and services that use AI will be trusted and adopted.

The consequence of this argument is that, if developing countries in Africa are to reap the rewards of adopting AI, innovation needs to be in the hands of those who understand the sociocultural factors that impact on trust, an understanding of which is essential for adoption and the realization of the benefits of the technological invention. In other words, as Michel Bézy puts it, it is imperative to "develop the African innovation market where new ICT solutions that are adapted to Africa's environment and needs will be developed by Africans for Africa" [16]. In the following section, we will ground this argument by looking more closely at the importance of AI to Africa.

3 AI in Africa

AI is having an increasingly positive impact in Africa, in sectors such as energy, healthcare, agriculture, public services, and financial services [68]. It has the potential to drive economic growth, development, and democratization, reducing poverty, improving education, supporting healthcare delivery, increasing food production, improving the capacity of existing road infrastructure by increasing traffic flow, improving public services, and improving the quality of life of people with disabilities [72].

In the energy sector, AI and internet of things (IoT) technology have been proactively supported financially and adopted. Companies such as Engie [26], the French electricity provider, have funded several energy tech startups such as Fenix International [30], Mobisol [63], PowerCorner [73], and partnered with Equatorial Power [27], among others. Through pay-as-you-go energy products in Uganda, Zambia,

Kenya, Tanzania, Rwanda, Nigeria, Benin, Ivory Coast, Côte d'Ivoire, and Mozambique, AI is being used to score users and predict demand, making the products more affordable and adaptable. For example, Fenix uses predictive analytics to extend energy services and products to those who previously lacked access. Because of solutions like these, more of the over 600 million people in Sub-Saharan Africa who lack electricity can now be connected [74]. Initially targeting East Africa, the Electricity Growth and Use in Developing Economies Initiative [25] is using data analytics and deep learning, drawing on a broad base of historic consumption data and satellite imagery, with the goal of providing an open electricity consumption growth prediction service for individual businesses and residences. It is also using daily night-time illumination satellite imagery to provide wide-area, long-term estimates of grid stability across Sub-Saharan Africa.

In the healthcare sector, AI helps address the shortage of doctors through telemedicine and access to medical supplies through drone delivery [15]. In Kenya and South Africa, Tambua Health is providing an AI-assisted handheld tool that uses deep learning algorithms to assess the health of the heart and lungs with minimal training and minimal wait-time [82]. The government of Rwanda recently announced plans to open a robotics cancer training centre [49] in collaboration with a France-based research institute for digestive cancer [48]. This centre will offer laparoscopic training and R&D for minimally-invasive surgery. It has already recruited software developers and research engineers from local universities, leveraging graduate education in robotics and computer vision. In conjunction with the United Nations Development Programme, Rwanda has also recently deployed smart anti-epidemic robots in their successful fight against COVID-19 [85]. The use of drones in healthcare also has significant potential. For example, Silicon Valley-startup Zipline [90] partnered with the Rwandan government to deliver more than 50 types of blood products to rural hospitals and clinics using custom-designed drones [24]. The Zipline drones have a range of more than 100 km and, as of 2018, 12,000 units of blood had been delivered on more than 6,000 flights. As soon as a drone leaves the launch catapult, it is fully autonomous [1]. More than 30 of its 100 employees are Rwandan. The authors of an IEEE Spectrum article sum up the Zipline operation up well: "In the distance, we can hear the faint buzz of another Zip returning home after making its delivery of blood. Anywhere else on Earth, it would be futuristic. In rural Rwanda, it's just routine" [1].[3]

In the agricultural sector, AI has the potential to improve productivity and efficiency at all stages of the value chain, allowing small-holder farmers to increase their income through higher crop yield and greater price control, detection and precision treatment of pests and diseases, monitoring soil condition and targeted deployment of fertilizer, creation of virtual cooperatives to aggregate crop yield, broker better prices, and exploit economies of scale. There are a few companies already deploying AI in agriculture. Gro Intelligence, for example, is an analytics company in Kenya which is filling a global gap in the world of agricultural data by deploying artificial

[3] To see a Zipline drone drop blood packs to a clinic in rural Rwanda, see https://bit.ly/2pfnB6l (time interval 0′ 52″–1′ 22″).

intelligence systems for food security and climate stability solutions [35]. The startup recently closed an $85 million Series B round.

Drone technology for precision agriculture—using targeted interventions that optimize the use of available resources to increase profitability and sustainability of agricultural operations [33]—is a potential game-changer for the African continent, albeit one that requires a skilled workforce with competencies ranging from planning flight itineraries, operating GIS and data analysis software, interpreting data, and providing agronomic advice [10]. Their use is growing quickly in situations where crops are grown as a monoculture on large holdings and local companies, e.g. [19] in Rwanda, [2, 42] in South Africa, are now addressing the challenges of deployment for small-scale, multi-crop farms. Significantly, this also opens up opportunities to develop systems that can automatically incorporate agronomic expertize to identify appropriate interventions based on real-time sensor data, e.g. soil moisture level, pH level, nitrate level, and temperature, exploiting IoT platforms [12]. Existing projects include the Third Eye project in Mozambique and Kenya using low-cost drones to provide advice to farmers on irrigation and when to apply fertilizer and sow seeds [83], resulting in an increase in crop production by 41%, a reduction of water use by 9%, and a 55% increase in water productivity [10]. Microsoft is working to apply its Farmbeats platform [86] in developing countries by lowering the high cost associated with dense deployment of sensors, exploiting sparsely distributed sensors and aerial imagery to generate precision maps, and using smart phones attached to hand-carried low-cost tethered helium balloons [81].

Drones are also used for surveying elephants in Burkina Faso [87], similarly as a tool to combat poaching of rhinos in South Africa [64], and for humanitarian aid, e.g. for detailed mapping and modelling flood risks in Dar es Salaam, Tanzania, Africa's fastest growing city where 70% of the people live in informal, unplanned settlements with inadequate infrastructure [80].

Evidently, the spirit of entrepreneurship and innovation in AI is flourishing in Africa. For example, Hepta Analytics, a startup by seven Carnegie Mellon University Africa graduates, specializes in helping local industry leverage the benefits of data science [37]. One of their products, Najua, focusses on using machine learning to make the web available in local African languages [65]. However, progress can be inhibited for so-called "low-resourced" languages, i.e. languages for which few digital or computational data resources exist [66], because of the lack of sufficient training data. This is symptomatic of a problem that is endemic to almost all applications of machine learning in Africa: the paucity of data. Another Carnegie Mellon University Africa graduate heads a team of entrepreneurs deploying IoT technology on tea plantations in Uganda [47]. Ubenwa is a mobile app developed by a start-up in Nigeria. It uses AI to analyse acoustic signatures in newborn babies to detect early signs on perinatal asphyxia, a leading cause of neonatal disability and death. Given that many developing countries in Africa have an agrarian economy driven primarily by smallholder farmers, agriculture, as we have noted already, is a popular target for African tech entrepreneurs. For example, uLima is a smartphone app for farmers, agro-dealers, and others in the sector. It provides access to crop and livestock management information, weather and market price information, as well as customized crop and livestock calendars, all focused on improving farm productivity and the livelihoods of farmers and their families [84]. Since most farmers in Africa are smallholders and don't have access to smartphone technology, other companies provide similar services using lower-tech feature phones [44].

4 Accelerating the Exploitation and Adoption of AI in Africa

Despite the positive outlook presented above, widespread benefits of AI won't materialize without appropriate investment, education, and a legal framework to safeguard ethical research, development, and innovation [20, 68] along with access to a deep pool of data that is relevant to Africa and initiatives to build trust [72]. The crucial importance of ensuring sufficient representative data is available cannot be overstated. One study targeting machine translation of over thirty African languages showed that participatory research offer a potential scalable solution to this problem [66].

Support from foreign agencies can help, particularly where it is collaborative and well-targeted. One example is the strategic partnership with the Smart Africa alliance [79] for a digital Africa, supported by the German Federal Ministry for Economic Cooperation and Development [23]. This program aims to advance Africa's development through digital innovations. The FAIR Forward—Artificial Intelligence for All program [29], also funded by the German government, seeks to strengthen local technical know-how on AI in Africa and Asia. To date, they have formed partnerships with five countries, four in Africa (Ghana, Rwanda, South Africa, and Uganda), and the fifth with India. Other programs target the promotion of tech start-ups, e.g. the Make-IT Initiative [58] and Google Startups Accelerator Africa [34].

5 The Downsides of AI

It's not all good news, though. The deployment of AI in developed countries can have a severe negative impact on developing countries due to the phenomenon known as premature deindustrialization [53, 75] which sees low-wage developing countries having fewer opportunities for industrialization before achieving income levels comparable to those in developed countries. Kenya, Nigeria, and South Africa, for example, are projected to have approximately 5.5%, 8.5%, and 12.5%, respectively, of their workforce displaced by automation [59]. A report by the Oxford Martin School at the University of Oxford and Citi summarizes the situation in Africa in stark terms [31]:

> In most of sub-Saharan Africa, the manufacturing share of output has persistently declined over the past 25 years. The share of jobs in manufacturing is even smaller: just over 6% of

all jobs. This figure barely changed over the course of the three decades leading up to 2008, while manufacturing employment in Asia grew from 11% to 16% over the same period.

Developing countries lose their competitive advantage in manufacturing due to the lower cost automation in developed countries and therefore miss out on the economic benefits that developed countries enjoyed as their workforces moved from low-value work to manufacturing before progressing to a post-industrial service economy. Consequently, developing countries are increasingly likely not to have the opportunity for rapid economic growth by shifting workers from farms to factory jobs because (a) automation undermines the labour cost advantage and (b) developments in robotics and additive manufacturing allow companies in advanced economies to locate production closer to domestic markets in automated factories, allowing this work to be moved closer to home in the developed countries.

AI and robotics can help offset this trend, at least to some extent [53]. This growing concern about premature deindustrialization in emerging and developing countries will require new growth models: "because skilled jobs are substantially less susceptible to automation, the best hope for developing and emerging economies alike is to upskill their workforce" [31].

It is also important to keep in mind that AI can be used for negative purposes, either intentionally or unintentionally, e.g., by fomenting religious, ethnic, social, and political divisions through deep fake misinformation [15], the lack of democratization in AI, including implicit and explicit bias in the data that are used to train the AI models.

For concrete examples of this danger, consider the following. Buolamwini et al.'s [18] evaluation of bias in automated facial analysis algorithms and datasets with respect to phenotypic subgroups showed an uneven distribution of skin colors in datasets. The datasets were overwhelmingly composed of lighter-skinned subjects with 79.6% for IJB-A and 86.2% for Adience. Upon the evaluation of three commercial gender classification systems (Microsoft, IBM, and Face++) using a corrected and representative dataset, results showed that darker-skinned females are the most mis-classified group with error rates of up to 34.7%. The maximum error rate for lighter-skinned males was 0.8%.

Wilson et al. [89] showed that object detection systems like those used in autonomous vehicles had uniformly poorer accuracy (5% less accurate) when detecting pedestrians with darker skin types. This is a serious issue for autonomous driving in Africa. Unfair decisions have also been experienced in search engines. Search algorithms (e.g. Google) have also been noted in [67] for reproducing and reinforcing social stereotyping and racism among the minority groups.

COMPAS is a proprietary tool used to assess the probability of recidivism [17]. However, there are concerns about COMPAS being discriminatory [36]. Larson et al. [54] disclosed that the COMPAS algorithm produces recidivism risk assessments that are attributed with higher false positive rate and much lower false negative rates for darker-skinned defendants compared to light-skinned defendants.

These instances have negative impact on the perception of trust and confidence of Africans, which in turn, affect the adoption of such technologies especially when they are not developed locally.

Despite all these potential downsides, AI still comes with promising advantages. In the guise of digital forensics and cybersecurity, it assists in identifying misinformation. It also assists in healthcare administration, business, and education. As we saw in Sect. 3, and as we will see in the next section, Africa can harness the many upsides of AI, especially through education.

6 AI Education in Africa Is the Key to Progress

In Sect. 2, we argued that Africa must participate in building and deploying its own AI if there is to be trust, adoption, and consequent benefits. In turn, this meaningful exploitation of AI requires a skilled workforce and the demand for AI engineers will increase significantly [72]. The challenge for the education system in Africa—at every level, from primary to third level and beyond—is to adapt to the new needs for Industry 4.0 by focusing in primary and secondary education on science, technology, engineering, and mathematics (STEM) subjects [13, 14], and in third- and fourth-level education on imparting the requisite skills that will allow graduates to adapt quickly to the evolving AI landscape [12].

There is evidence that this is happening. For example, with sponsorship from Google and Facebook, the African Institute of Mathematical Sciences [9] launched a Master of Machine Intelligence in Kigali, Rwanda, in 2018 [8, 20]. With the backing of the Government of Rwanda, Carnegie Mellon University Africa offers two Masters programs, one in in Electrical and Computer Engineering and one in Information Technology, both targeting key skills in AI, machine learning, data science, software engineering, cyber-security, telecommunications, and energy systems [21], with a dedicated Master program in Engineering AI in the pipeline. With the support of over US$10 million from the African Development Bank, the Rwandan Development Board completed the construction of a new campus for CMU-Africa in 2019 as part of Kigali Innovation City (KIC) [12]. Google opened an AI Research Lab in Accra, Ghana, in 2018; and the Deep Learning Indaba summer schools [45] attract more applicants than they can accept from almost half the countries in Africa [20]. When the constraints of the Covid-19 pandemic meant that they could not hold these summer schools, the organizers, still committed to their mission to "strengthen African machine learning and artificial intelligence", launched instead a program of AI4D innovation grants [46]. The successful annual IBRO-SIMONS Computational Neuroscience Imbizo[4] also had to be suspended during the pandemic but will be held again in 2022 [43].

The signs are positive at secondary and primary level too. For example, 42 teams from 18 countries in Africa attended the fourth edition of the Pan-African

[4] *Imbizo* is a Xhosa word meaning "a gathering to share knowledge" [43].

Robotics Competition (PARC) in Ghana in 2019 [71]. In 2018, senior students at the Massachusetts Institute of Technology helped organize a three-week robotics camp focusing on agriculture, bringing together more than 40 students aged between 14 and 17 years (18 boys and 22 girls) drawn from 20 schools in Rwanda [62]. Fundi Bots [32] in Uganda provides classes in robotics, and has already reached 10,000 children in an effort to encourage them to learn STEM subjects in a fun, interactive, and practical manner. In addition to that, they launched a special program, Fundi Girls to increase the significant under-representation of girls and young women in STEM education. Start-up companies such as the Children's Creativity Lab [22] in Rwanda are also looking to cater for younger children.

The plenary talk by Ayorkor Korsah, Ashesi University, Ghana, at ICRA 2015 on robotics in education in Africa [52] highlighted the relevance of robotics to Africa. She surveyed the various activities in promoting robotics in, e.g., Ghana, Kenya, South Africa, and Egypt, and emphasized the need to find ways to empower young Africans to provide robotics solutions to African problems.

The availability of a good education is not enough, however: that education must also be accessible. Regrettably, most young Africans cannot afford the high cost of education and it falls to governments to put in place free or reasonable-cost education and scholarships. Again, there is growing evidence that this is beginning to happen. For example, the Government of Rwanda provides the financial backing to reduce the costs of the CMU-Africa Masters programs to place them within the reach of students across Africa, with additional scholarships provided by, for example, Mastercard [60], Smart Africa Smart Africa [79], and the Mandela Institute for Development Studies [61], helping to bridge the remaining shortfall in required funding. With the backing of Google and Facebook, the AIMS African Master of Machine Intelligence is fully funded [8].

In a testimony to the drive and motivation to engage with AI in Africa, young Africans are taking matters of education in AI into their own hands. For example, young women and men meet every week to teach each other the latest techniques in AI, machine learning, and data science under the guise of AI Saturdays Lagos in Nigeria [6]. Some of these young people are students, others are budding entrepreneurs, while others are waiting (usually successfully) to be admitted to a Ph.D. program abroad, with the clear intention of returning to take up where they left off after they have gained the extra knowledge and know-how that will enable them to help transform the educational and technological landscape in Africa. This will foster the design and deployment of trustworthy AI technologies by Africans for Africans.

7 Conclusions

There are many factors that need to be addressed to achieve the benefits of AI in Africa, but success hinges not on enabling Africa, but on facilitating Africa's nascent efforts to enable itself. As Moustapha Cissé, head of the Google AI Centre in Ghana, states, "Fewer African AI researchers and engineers means fewer opportunities to use AI to improve the lives of Africans" [20].

In that context, it is important to keep in mind that the relevance of AI to Africa is not just to exploit it for social and economic development. There are several reasons why Africa can also play an important role in advancing the discipline of AI. For example, the elimination of bias and implicit discrimination in AI algorithms can be addressed in many ways. One of these is to increase the diversity of AI developers by growing the number of AI researchers and innovators in Africa.

So, how can the global AI community play its part in facilitating Africa's effort to enable itself? Greater attention can be paid to ensuring training sets for machine learning algorithms are available and that they are not biased and reflect the social, cultural, and ethnographic reality of African people, including the wide diversity of local languages. National and international research programs can allow African researchers to participate in collaborative research projects and provide funding for marginal costs. Educators can take sabbatical leave to teach in African universities. Researchers can do the same. Authors and instructors linked with online courses, such as the ones that AI Saturdays Lagos use to bootstrap their education, can give targeted guest lectures, either in person or remotely. GPU cloud providers can make their platforms accessible free of charge. Universities can greatly increase the opportunities for young aspiring researchers and developers to pursue fully-funded Masters and Ph.D. degrees in AI, machine learning, robotics, data science, and cybersecurity. All of this will help bright, ambitious, innovative African people to play their pivotal role in Africa's continuing efforts to embrace trustworthy AI, leading to the creation and adoption of African solutions to African problems.

Acknowledgements The ideas and arguments put forward in this chapter can be traced to many sources. However, they were heavily influenced by our collective experience at Carnegie Mellon University Africa, Kigali, Rwanda. We would like to thank everyone there for making it such a special place to learn and teach. Daniel Omeiza gratefully acknowledges support by the UK's Engineering and Physical Sciences Research Council (EPSRC) through project RoboTIPS: Developing Responsible Robots for the Digital Economy, grant reference EP/S005099/1. Charity Delmus Alupo gratefully acknowledges support by Science Foundation Ireland through the SFI Centre for Research Training in Machine Learning, grant reference 18/CRT/6183.

References

1. Ackerman E, Koziol M (2019) The blood is here: Zipline's medical delivery drones are changing the game in Rwanda. IEEE Spectr 56(5):24–31
2. Aerobotics (2021) https://www.aerobotics.com/press
3. AI EU (2021) European Commission white paper on artificial intelligence—a European approach to excellence and trust. https://ec.europa.eu/info/sites/info/files/commissionwhite-paper-artificial-intelligence-feb2020en.pdf
4. AI Japan (2021) OECD AI in Japan. https://oecd.ai/dashboards/countries/Japan

5. AI Rwanda (2021) The future society—development of Rwanda's national artificial intelligence policy. https://thefuturesociety.org/2020/08/31/development-of-rwandasnational-artificial-intelligence-policy/
6. AI Saturdays (2021) Lagos. Tech rep. http://www.aisaturdayslagos.com/index.html
7. AI UAE (2021) Artificial intelligence strategy 2031. http://www.uaeai.ae/en/
8. AIMMI (2021) African Institute for Mathematical Sciences (AIMS) master's in machine intelligence. https://aimsammi.org/
9. AIMS (2021) The African institute for mathematical sciences. https://aims.ac.rw/
10. AU-NEPAD (2018) Drones on the horizon—transforming Africa's agriculture. Tech. rep., The African Union and NEPAD, High-Level African Pandel on Emerging Technologies
11. Babalola SF, Alokan OA (2013) African concept of time, a socio-cultural reality in the process of change. J Educ Pract 4(7):143–147
12. Badiane O, Braun Jv (2019) Byte by byte: policy innovation for transforming Africa's food system with digital technologies. Tech. rep., Malabo Montpelier Panel
13. Banga K, te Velde DW (2018) Digitalisation and the future of African manufacturing. Tech. rep., Supporting Economic Transformation
14. Banga K, te Velde DW (2018) Digitalisation and the future of African manufacturing: briefing paper. Tech. rep., Supporting Economic Transformation
15. Besaw C, Filitz J (2019) AI & global governance: AI in Africa is a double-edged sword. Tech. rep., United Nations University. https://cpr.unu.edu/ai-in-africa-is-a-doubleedged-sword.html
16. Bézy (2021) African Oye. http://brel54.blogspot.com/
17. Brennan T, Dieterich W (2018) Correctional offender management profiles for alternative sanctions (COMPAS). Handbook of recidivism risk/needs assessment tools, p 49
18. Buolamwini J, Gebru T (2018) Gender shades: intersectional accuracy disparities in commercial gender classification. In: Conference on fairness, accountability and transparency, pp 77–91
19. Charis (2021) Charis unmanned aerial solutions. https://charisuas.com/
20. Cisse M (2018) Look to Africa to advance artificial intelligence. Nature 562:461
21. CMU-Africa (2021) Carnegie Mellon University Africa masters programs in electrical and computer engineering and in information technology. https://www.africa.engineering.cmu.edu
22. Creativity (2021) Children's creativity lab. http://creativity.rw/robotics
23. Digital Africa (2021) Smart Africa—alliance for a digital Africa. https://toolkitdigitalisierung.de/en/smart-africa-eine-allianz-fuer-ein-digitales-afrika/
24. Duan N (2018) On-demand medical drone delivery. Stanf Soc Innov Rev 16(4):6–7
25. e-GUIDE (2021) Electricity growth and use in developing economies. https://eguide.io/
26. ENGIE (2021) https://www.engie.com/en
27. Equatorial Power (2021) http://equatorial-power.com/
28. EUAI (2021) European Commission ethics guidelines for trustworthy AI. https://ec.europa.eu/futurium/en/ai-alliance-consultation
29. FAIR Forward (2021) Artificial intelligence for all. https://toolkitdigitalisierung.de/en/fair-forward/
30. Fenix (2021) https://www.fenixintl.com/
31. Frey C, Osborne M, Holmes C, Rahbari E, Curmi E, Garlick R, Chua J, Friedlander G, Chalif P, McDonald G, Wilkie M (2016) Technology at work v2.0: the future is not what it used to be. White paper, Oxford Martin School, University of Oxford, and Citigroup
32. Fundi Bots (2021) https://fundibots.org/fundi-bots-classes/
33. Gebbers R, Adamchuk VI (2010) Precision agriculture and food security. Science 327
34. Google Startup (2021) Google startup accelerators Africa. developers.google.com/community/accelerators
35. Gro Intelligence (2021) https://gro-intelligence.com/
36. Hamilton M (2019) The sexist algorithm. Behav Sci Law 37(2):145–157
37. Hepta Analytics (2021) www.heptanalytics.com/
38. Hofman LJ, Lawson-Jenkins K, Blum J (2006) Trust beyond security: an expanded trust model. Commun ACM 49(7)

39. Hofstede G (1980) Cultures consequences: international differences in work-related values. Sage Publications
40. Hofstede G (1991) Cultures and organizations: software of the mind. McGraw-Hill
41. Hofstede G, Hofstede GJ, Minkov M (2010) Cultures and organizations: software of the mind. McGraw-Hill
42. IAS (2021) Integrated aerial systems. https://iasystems.co.za/
43. IBRO-SIMONS (2021) Computational Neuroscience Imbizo. https://imbizo.africa/
44. iCow (2021) https://www.icow.co.ke/
45. Indaba (2021) Deep learning Indaba. https://deeplearningindaba.com/2020/
46. IndabaX (2021) Deep learning Indaba AI4D innovation grants. https://deeplearningindaba.com/blog/2020/09/recipients-of-the-2020-indabax-ai4dinnovation-grants/
47. InterConnectPoint (2021) IoT application enhancement and remote asset management. https://iconnectpoint.com/
48. IRCAD (2021) Laproscopic training center for minimally invasive surgery. https://www.ircad.fr/
49. IRCAD (2021) Laproscopic training center for minimally invasive surgery in Africa. http://www.ircad.africa
50. Kelly JE (2015) Computing, cognition and the future of knowing. IBM Corporation, White paper
51. Köhler H (2014) On the impossibility of speaking of Africa. In: German-African Cooperation in Education and Research BMBF Africa Days, Federal Ministry of Education and Research, pp 27–31
52. Korsah A (2021) Plenary talk at ICRA 2015 on robotics in education Africa. https://resourcecenter.ieee-ras.org/conferences-videos/icra-videos/824490.htm
53. Kozul-Wright R (2016) Robots and industrialization in developing countries. Tech. rep., United Nations Conference on Trade and Development (UNCTAD)
54. Larson J, Mattu S, Kirchner L, Angwin J (2016) How we analyzed the COMPAS recidivism algorithm. ProPublica 9(1)
55. Lee JD, See KA (2004) Trust in automation: designing for appropriate reliance. Hum Factors 46(1):50–80
56. Licklider JCR (1960) Man-computer symbiosis. IRE Trans Human Factors Electron HFE-1:4–11
57. Lum K, Isaac W (2016) To predict and serve? Significance 13(5):14–19
58. Make-IT (2021) https://toolkit-digitalisierung.de/make-it-initiative/
59. Manyika J, Lund S, Chui M, Bughin J, Woetzel J, Batra P, Ko R, Sanghvi S (2017) Jobs lost, jobs gained: workforce transitions in a time of automation. Tech. rep., McKinsey Global Institute
60. Mastercard Foundation (2021) https://mastercardfdn.org/
61. MINDS (2021) The Mandela Institute for development studies. https://minds-africa.org/
62. MIT (2021) Robotics Camp, Kigali, Rwanda. http://africa.mit.edu/news-andevents/robotics-camp-ends-with-call-for-more-investment/
63. Mobisol (2021) https://plugintheworld.com/
64. Mulero-Pázmány M, Stolper R, Essen Lv, Negro J, Sassen T (2014) Remotely piloted aircraft systems as a rhinoceros anti-poaching tool in Africa. PLoS ONE
65. Najua (2021) Say hello to your new multilingual assistant. http://translate.najua.ai/
66. Nekoto W, Marivate V, Matsila T, Fasubaa T, Fagbohungbe T, Akinola SO, Muhammad S, Kabongo Kabenamualu S, Osei S, Sackey F, Niyongabo RA, Macharm R, Ogayo P, Ahia O, Berhe MM, Adeyemi M, Mokgesi-Selinga M, Okegbemi L, Martinus L, Tajudeen K, Degila K, Ogueji K, Siminyu K, Kreutzer J, Webster J, Ali JT, Abbott J, Orife I, Ezeani I, Dangana IA, Kamper H, Elsahar H, Duru G, Kioko G, Espoir M, van Biljon E, Whitenack D, Onyefuluchi C, Emezue CC, Dossou BFP, Sibanda B, Bassey B, Olabiyi A, Ramkilowan A, Öktem A, Akinfaderin A, Bashir A (2020) Participatory research for low-resourced machine translation: a case study in African languages. In: Findings of the Association for Computational Linguistics: EMNLP 2020. Association for Computational Linguistics, pp 2144–2160

67. Noble SU (2018) Algorithms of oppression: how search engines reinforce racism. NYU Press
68. Novitske L (2018) The AI invasion is coming to Africa (and it's a good thing). Stanf Soc Innov Rev
69. OECDAI (2021) National AI policies & strategies. https://oecd.ai/dashboards
70. One World Nations (2019) Official and spoken languages of African countries. https://www.nationsonline.org/oneworld/africanlanguages.htm
71. PARC (2019) Pan-African robotics competition. http://parcrobotics.org/
72. Pillay N, Partnership A (2018) Artificial intelligence for Africa: an opportunity for growth, development, and democratization. University of Pretoria, White paper
73. PowerCorner (2021) https://engie-powercorner.com/en/
74. Quartz Africa (2021) https://rb.gy/qwta6q
75. Rodrik D (2016) Premature deindustrialization. J Econ. Growth 21(1):1–33
76. Rose J (2010) Software innovation: eight work-style heuristics for creative software developers. Software Innovation, Dept. of Computer Science, Aalborg University
77. Schwab K (2021) The fourth industrial revolution: what it means, how to respond, Word Economic Forum. https://www.weforum.org/agenda/2016/01/the-fourth-industrialrevolution-what-it-means-and-how-to-respond/
78. Shanahan M (2015) The technological singularity. MIT Press
79. Smart Africa (2021) https://smartafrica.org/
80. Soesilo D, Meier P, Lessard-Fontaine A, Du Plessis J, Stuhlberger C, Fabbroni V (2016) Drones in humanitarian action. Tech. rep., Retrieved from Drones for Humanitarian and Environmental Applications: https://goo.gl/aDtz4p
81. Swamy AN, Kumar A, Patil R, Jain A, Kapetanovic Z, Sharma R, Vasisht D, Swaminathan M, Chandra R, Badam A, Ranade G, Sinha S, Nambi A (2019) Low-cost aerial imaging for small holder farmers. In: Proceedings of the 2nd ACM SIGCAS conference on computing and sustainable societies (COMPASS), Accra, Ghana
82. Tambua Health (2021) Precise, non-invasive and radiation free lung imaging using acoustic technology + machine learning. https://www.tambuahealth.com
83. Third Eye (2021) http://www.thirdeyewater.com/
84. uLima (2021) http://ulima.co/
85. UNDP (2021) United Nations Development Programme and government of Rwanda deploy smart anti-epidemic robots to fight against COVID-19!. https://www.africa.undp.org/content/rba/en/home/presscenter/articles/2020/undp-deployssmart-anti–epidemic-robots-to-fight-against-covid.html
86. Vasisht D, Kapetanovic Z, Won J, Jin X, Chandra R, Sinha SN, Kapoor A, Sudarshan M, Stratman S (2017) Farmbeats: an IoT platform for data-driven agriculture. In: Proceedings of the 14th USENIX symposium on networked systems design and implementation, Boston, MA, USA, pp 515—529
87. Vermeulen C, Lejeune P, Lisein J, Sawadogo P, Bouché P (2013) Unmanned aerial survey of elephants. PLoS ONE 8(2)
88. Vernon D (2019) Robotics and artificial intelligence in Africa. IEEE Robot Autom Mag 26(4):131–135
89. Wilson B, Hoffman J, Morgenstern J (2019) Predictive inequity in object detection. arXiv:190211097
90. Zipline (2021) https://flyzipline.com/

Index

© The Editor(s) (if applicable) and The Author(s), under exclusive license to Springer Nature Switzerland AG 2022

M. I. A. Ferreira and M. O. Tokhi (eds.), *Towards Trustworthy Artificial Intelligent Systems*, Intelligent Systems, Control and Automation: Science and Engineering 102,
https://doi.org/10.1007/978-3-031-09823-9

The manufacturer's authorised representative in the EU is Springer Nature Customer Service Centre GmbH, Europaplatz 3, 69115 Heidelberg, Germany. If you have any concerns regarding our products, please contact ProductSafety@springernature.com

Printed and bound by CPI Group (UK) Ltd, Croydon, CR0 4YY
12/08/2026
02194534-0001